AFRICA

and the

MIDDLE EAST

A Continental Overview
of Environmental Issues

THE WORLD'S ENVIRONMENTS

KEVIN HILLSTROM AND
LAURIE COLLIER HILLSTROM, SERIES EDITORS

Global warming, rain forest destruction, mass extinction, overpopulation—the environmental problems facing our planet are immense and complex.

ABC-CLIO's series The World's Environments offers students and general readers a handle on the key issues, events, and people.

The six titles in the series examine the unique—and common—problems facing the environments of every continent on Earth and the ingenious ways local people are attempting to address them. Titles in this series:

Africa and the Middle East

Asia

Australia, Oceania, and Antarctica

Europe

Latin America and the Caribbean

North America

AFRICA

and the

MIDDLE EAST

A Continental Overview
of Environmental Issues

KEVIN HILLSTROM
LAURIE COLLIER HILLSTROM

A B C 🌊 C L I O
Santa Barbara, California
Denver, Colorado   Oxford, England

363.70096
H655a

Copyright © 2003 by Kevin Hillstrom and Laurie Collier Hillstrom

All rights reserved. No part of this publication may be reproduced, stored in a retrieval system, or transmitted, in any form or by any means, electronic, mechanical, photocopying, recording, or otherwise, except for the inclusion of brief quotations in a review, without prior permission in writing from the publishers.

Library of Congress Cataloging-in-Publication Data

Hillstrom, Kevin, 1963–
 Africa and the Middle East : a continental overview of environmental issues / Kevin Hillstrom, Laurie Collier Hillstrom.
     p. cm. — (The world's environments)
   Includes index.
   ISBN 1-57607-688-1 (hardcover : alk. paper) ISBN 1-57607-693-8 (e-book)
   1. Africa—Environmental conditions. 2. Environmental sciences—Africa. 3. Middle East—Environmental conditions. 4. Environmental sciences—Middle East. I. Hillstrom, Laurie Collier, 1965– II. Title.

  GE160.A35H55 2003
  363.7'0096—dc22   2003020747

07 06 05 04 03   10 9 8 7 6 5 4 3 2 1

This book is also available on the World Wide Web as an eBook. Visit http://www.abc-clio.com for details.

ABC-CLIO, Inc.
130 Cremona Drive, P.O. Box 1911
Santa Barbara, California 93116–1911

This book is printed on acid-free paper ∞ .
Manufactured in the United States of America

2/04

Contents

List of
Tables and Figures

Tables

Figures

Introduction

As the nations of the world enter the twenty-first century, they confront a host of environmental issues that demand attention. Some of these issues—pollution of freshwater and marine resources, degradation of wildlife habitat, escalating human population densities that place crushing demands on finite environmental resources—have troubled the world for generations, and they continue to defy easy solutions. Other issues—global climate change, the potential risks and rewards of genetically modified crops and other organisms, unsustainable consumption of freshwater resources—are of more recent vintage. Together, these issues pose a formidable challenge to our hopes of building a prosperous world community in the new millennium, especially since environmental protection remains a low priority in many countries. But despite an abundance of troubling environmental indicators, positive steps are being taken at the local, regional, national, and international levels to implement new models of environmental stewardship that strike an appropriate balance between economic advancement and resource protection. In some places, these efforts have achieved striking success. There is reason to hope that this new vision of environmental sustainability will take root all around the globe in the coming years.

The World's Environments series is a general reference resource that provides a comprehensive assessment of our progress to date in meeting the numerous environmental challenges of the twenty-first century. It offers detailed, current information on vital environmental trends and issues facing nations around the globe. The series consists of six volumes, each of which addresses conservation issues and the state of the environment in a specific region of the world: individual volumes for *Asia*, *Europe*, and *North America*, published in spring 2003, will be joined by *Africa and the Middle East*; *Australia, Oceania, and Antarctica*; and *Latin America and the Caribbean* in the fall of the same year.

Each volume of The World's Environments includes coverage of issues unique to that region of the world in such realms as habitat destruction, water pollution, depletion of natural resources, energy consumption, and development. In addition, each volume provides an overview of the region's response to environmental matters of worldwide concern, such as global warming. Information on these complex issues is presented in a manner that is informative, interesting, and understandable to a general readership. Moreover, each book in the series has been produced with an emphasis on objectivity and utilization of the latest environmental data from government agencies, nongovernmental organizations (NGOs), and international environmental research agencies, such as the various research branches of the United Nations.

Organization

Each of the six volumes of The World's Environments consists of ten chapters devoted to the following major environmental issues:

Population and Land Use. This chapter includes continental population trends, socioeconomic background of the populace, prevailing consumption patterns, and development and sprawl issues.

Biodiversity. This chapter reports on the status of flora and fauna and the habitat upon which they depend for survival. Areas of coverage include the impact of alien species on native plants and animals, the consequences of deforestation and other forms of habitat degradation, and the effects of the international wildlife trade.

Parks, Preserves, and Protected Areas. This chapter describes the size, status, and biological richness of area park systems, preserves, and wilderness areas and their importance to regional biodiversity.

Forests. Issues covered in this chapter include the extent and status of forest resources, the importance of forestland as habitat, and prevailing forest management practices.

Agriculture. This chapter is devoted to dominant farming practices and their impact on local, regional, and national ecosystems. Subjects of special significance in this chapter include levels of freshwater consumption for irrigation, farming policies, reliance on and attitudes toward genetically modified foods, and ranching.

Freshwater. This chapter provides detailed coverage of the ecological health of rivers, lakes, and groundwater resources, extending special attention to pollution and consumption issues.

Oceans and Coastal Areas. This chapter explores the ecological health of continental marine areas. Principal areas of coverage include the current state of (and projected outlook for) area fisheries, coral reef conservation, coastal habitat loss from development and erosion, and water quality trends in estuaries and other coastal regions.

Energy and Transportation. This chapter assesses historic and emerging trends in regional energy use and transportation, with an emphasis on the environmental and economic benefits and drawbacks associated with energy sources ranging from fossil fuels to nuclear power to renewable technologies.

Air Quality and the Atmosphere. This chapter reports on the current state of and future outlook for air quality in the region under discussion. Areas of discussion include emissions responsible for air pollution problems like acid rain and smog, as well as analysis of regional contributions to global warming and ozone loss.

Environmental Activism. This chapter provides a summary of the history of environmental activism in the region under discussion.

In addition, each volume of The World's Environments contains sidebars that provide readers with information on key individuals, organizations, projects, events, and controversies associated with specific environmental issues. By focusing attention on specific environmental "flashpoints"—the status of a single threatened species, the future of a specific wilderness area targeted for oil exploration, the struggles of a single village to adopt environmentally sustainable farming practices—many of these sidebars also shed light on larger environmental issues. The text of each volume is followed by an appendix of environmental and developmental agencies and organizations on the World Wide Web. Finally, each volume includes a general index containing citations to issues, events, and people discussed in the book, as well as supplemental tables, graphs, charts, maps, and photographs.

Coverage by Geographic Region

Each of the six volumes of The World's Environments focuses on a single region of the world: Africa and the Middle East; Asia; Australia, Oceania, and Antarctica; Europe; Latin America and the Caribbean; and North America. In most instances, the arrangement of coverage within these volumes was obvious, in accordance with widely recognized geographic divisions. But placement of a few countries was more problematic. Mexico, for instance, is recognized both as part of North America and as the northernmost state in Latin America.

Moreover, some international environmental research agencies (both governmental and nongovernmental) place data on Mexico under the North American umbrella, while others classify it among Central American and Caribbean nations. We ultimately decided to place Mexico in the Latin America volume, which covers Central and South America, in recognition of its significant social, economic, climatic, and environmental commonalities with those regions.

Similarly, environmental data on the vast Russian Federation, which sprawls over northern reaches of both Europe and Asia, is sometimes found in resources on Asia, and at other times in assessments of Europe's environment. Since most of Russia's population is located in the western end of its territory, we decided to cover the country's environmental issues in The World's Environments Europe volume, though occasional references to environmental conditions in the Russian Far East do appear in the Asia volume.

Finally, we decided to expand coverage in the Africa volume to cover environmental issues of the Middle East—also sometimes known as West Asia. This decision was made partly out of a recognition that the nations of Africa and the Middle East share many of the same environmental challenges—extremely limited freshwater supplies, for instance—and partly because of the space required in the Asia volume to fully explicate the multitude of grave environmental problems confronting Asia's central, southern, and eastern reaches. Coverage of other nations that straddle continental boundaries—such as the countries of the Caucasus region—are also concentrated in one volume, though references to some nations may appear elsewhere in the series.

Following is an internal breakdown of the volume-by-volume coverage for The World's Environments. This is followed in turn by overview maps for the current volume showing country locations and key cities and indicating physical features.

Africa and the Middle East

Middle East and North Africa:
Algeria
Bahrain
Cyprus
Egypt
Gaza
Iraq
Israel
Jordan
Kuwait
Lebanon
Libya
Morocco
Oman
Qatar
Saudi Arabia
Syrian Arab Republic
Tunisia
Turkey
United Arab Emirates
West Bank
Yemen

Sub-Saharan Africa:
Angola
Benin
Botswana
Burkina Faso
Burundi
Cameroon
Central African Republic
Chad
Congo, Democratic Republic of
  (Zaire)
Congo, Republic of the

Côte d'Ivoire
Equatorial Guinea
Eritrea
Ethiopia
Gabon
Gambia
Ghana
Guinea
Guinea-Bissau
Kenya
Lesotho
Liberia
Madagascar
Malawi
Mali
Mauritania
Mozambique
Namibia
Niger
Nigeria
Rwanda
Senegal
Sierra Leone
Somalia
South Africa
Sudan
Tanzania
Togo
Uganda
Zambia
Zimbabwe

Asia
Afghanistan
Armenia
Azerbaijan

Bangladesh
Bhutan
Cambodia
China
Georgia
India
Indonesia
Iran
Japan
Kazakhstan
Korea, Democratic People's
    Republic of (North)
Korea, Republic of (South)
Kyrgyzstan
Lao People's Democratic Republic
Malaysia
Mongolia
Myanmar (Burma)
Nepal
Pakistan
Philippines
Singapore
Sri Lanka
Tajikistan
Thailand
Turkmenistan
Uzbekistan
Vietnam

Australia, Oceania, and Antarctica
Australia
Cook Islands
Fiji
French Polynesia
Guam
Kiribati

Nauru
New Caledonia
Northern Mariana Islands
Marshall Islands
Micronesia, Federated States of
New Guinea
New Zealand
Palau
Papua New Guinea
Pitcairn Island
Samoa
Solomon Islands
Tonga
Tuvalu
Vanuatu
Wallis and Futuna
Various territories
*(Note: Antarctica is discussed in a
    stand-alone chapter)*

Europe
Albania
Austria
Belarus
Belgium
Bosnia and Herzegovina
Bulgaria
Croatia
Czech Republic
Denmark
Estonia
Finland
France
Germany
Greece
Hungary

Iceland
Ireland
Italy
Latvia
Lithuania
Macedonia, Republic of
Moldova
Netherlands
Norway
Poland
Portugal
Romania
Russian Federation
Slovakia
Slovenia
Spain
Sweden
Switzerland
Ukraine
United Kingdom
Yugoslavia

**Latin America
  and the Caribbean**
Argentina
Belize
Bolivia

Brazil
Caribbean territories
Chile
Colombia
Costa Rica
Cuba
Dominican Republic
Ecuador
El Salvador
Guatemala
Guyana
Haiti
Honduras
Jamaica
Mexico
Nicaragua
Panama
Paraguay
Peru
Suriname
Trinidad and Tobago
Uruguay
Venezuela

North America
Canada
United States

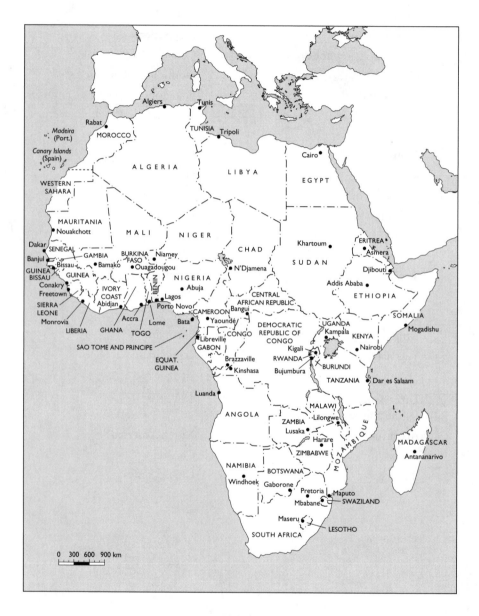

AFRICA

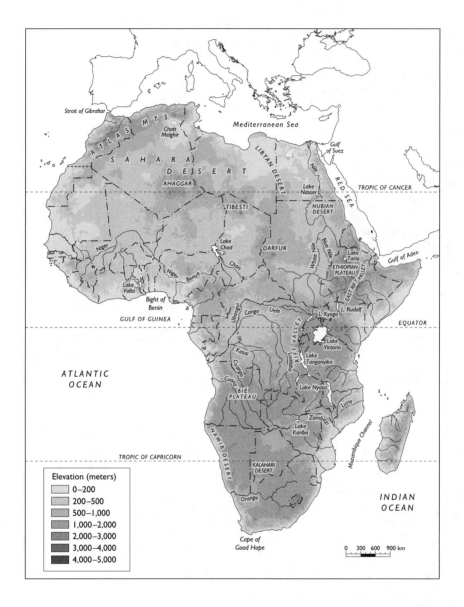

Strait of Gibraltar

Mediterranean Sea

Gulf
of Suez

ATLAS MTS.

Chott
Melghir

S A H A R A

D E S E R T

AHAGGAR

LIBYAN DESERT

Nile

RED SEA

Lake
Nasser

TROPIC OF CANCER

TIBESTI

NUBIAN
DESERT

Niger

Lake
Chad

DARFUR

White Nile

Blue Nile

Lake
Tana

Gulf of Aden

Niger

Benue

Chari

ETHIOPIAN
PLATEAU

EAST RIFT VALLEY

Lake
Volta

Bight of
Benin

GULF OF GUINEA

Ubangui

Congo

Uele

L. Kyoga

L. Rudolf

EQUATOR

RIFT VALLEY

Lake
Victoria

ATLANTIC
OCEAN

Kasai

Cuango

Lake
Tanganyika

Cuanza

BIÉ
PLATEAU

Lake Nyasa

Luria

Zambezi

Lake
Kariba

Mozambique Channel

TROPIC OF CAPRICORN

NAMIB DESERT

KALAHARI
DESERT

INDIAN
OCEAN

Orange

Cape of
Good Hope

Elevation (meters)

0–200
200–500
500–1,000
1,000–2,000
2,000–3,000
3,000–4,000
4,000–5,000

0 300 600 900 km

AFRICA

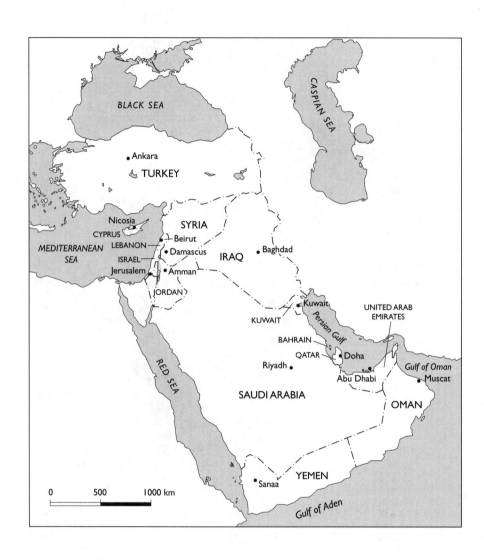

MIDDLE EAST

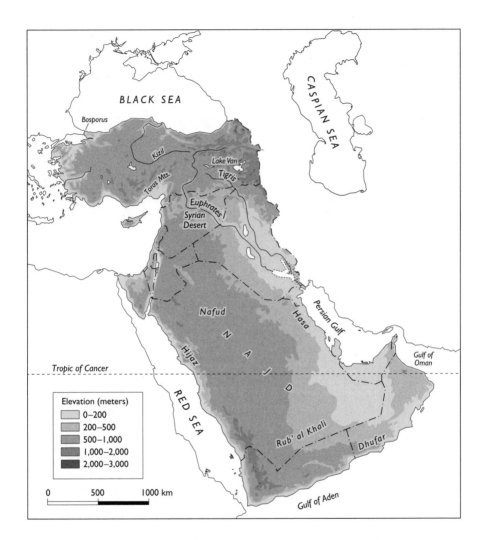

MIDDLE EAST

Acknowledgments

The authors are indebted to many members of the ABC-CLIO family for their fine work on this series. Special thanks are due to Vicky Speck, Martha Whitt, and Kevin Downing. We would also like to extend special thanks to our advisory board members, whose painstaking reviews played a significant role in shaping the final content of each volume, and to the contributors who lent their expertise and talent to this project.

Biographical Notes

Authors

KEVIN HILLSTROM and **LAURIE HILLSTROM** have written and edited award-winning reference books on a wide range of subjects, including American history, international environmental issues, environmental activism, outdoor travel, and business and industry. Works produced by the Hillstroms include *Environmental Leaders 1* and *2* (1997 and 2000), the four-volume *American Civil War Reference Library* (2000), the four-volume *Vietnam War Reference Library* (2000), *Paddling Michigan* (2001), *Encyclopedia of Small Business, 2d ed.* (2001), and *The Vietnam Experience: A Concise Encyclopedia of American Literature, Films, and Songs* (1998).

Advisory Board

J. DAVID ALLAN received his B.Sc. (1966) from the University of British Columbia and his Ph.D. (1971) from the University of Michigan. He served on the Zoology faculty of the University of Maryland until 1990, when he moved to the University of Michigan, where he currently is Professor of Conservation Biology and Ecosystem Management in the School of Natural Resources and Environment. Dr. Allan specializes in the ecology and conservation of rivers. He is the author of *Stream Ecology* (1995) and coauthor (with C. E. Cushing) of *Streams: Their Ecology and Life* (2001). He has published extensively on topics in community ecology and the influence of land-use on the ecological integrity of rivers. He serves or has served on committees for the North American Benthological Society, the Ecological Society of America, and the American Society of Limnology and Oceanography. He serves or has served on the editorial board of the scientific journals *Freshwater Biology* and *Journal of the North American Benthological Society,* and on scientific advisory committees for the American Rivers and Nature Conservancy organizations.

DAVID LEONARD DOWNIE is Director of Education Partnerships for the Earth Institute at Columbia University, where he has conducted research and taught courses on international environmental politics since 1994. Educated at Duke University and the University of North Carolina, Dr. Downie is the author of numerous scholarly publications on the Stockholm Convention, the Montreal Protocol, the UN Environment Program, and other topics in global environmental politics. From 1994 to 1999, Dr. Downie served as Director of Environmental Policy Studies at the School of International and Public Affairs, Columbia University.

CHRIS MAGIN was educated at Cambridge University, England. He took an undergraduate degree in Natural Sciences and a Ph.D. in Zoology, conducting fieldwork on hyraxes in Serengeti National Park, Tanzania. Since then he has been a professional conservationist, employed by various international organizations, mainly in Africa and Asia. He currently works for Flora and Fauna International. His special areas of interest are desert ungulates, ornithology, and protected area management.

JEFFREY A. MCNEELY is Chief Scientist at IUCN-The World Conservation Union, where he has worked since 1980. Prior to going to IUCN, he spent three years in Indonesia, two years in Nepal, and seven years in Thailand, working on various biodiversity-related topics. He has published over thirty books, including *Mammals of Thailand* (1975); *Wildlife Management in Southeast Asia* (1978); *National Parks, Conservation and Development: The Role of Protected Areas in Sustaining Society* (1984); *Soul of the Tiger* (1985); *People and Protected Areas in the Hindu Kush-Himalaya* (1985); *Economics and Biological Diversity* (1988); *Parks for Life* (1993); *Expanding Partnerships for Conservation* (1995); *Biodiversity Conservation in the Asia and Pacific Region* (1995); *A Threat to Life: The Impact of Climate Change on Japan's Biodiversity* (2000); *The Great Reshuffling: The Human Dimensions of Invasive Alien Species* (2001); and *Ecoagriculture: Strategies to Feed the World and Save Wild Biodiversity* (2003). He is currently working on a book on war and biodiversity. He was Secretary General of the 1992 World Congress on Protected Areas (Caracas, Venezuela), and has been deeply involved in the development of the Convention on Biological Diversity. He is on the editorial board of seven international journals.

CARMEN REVENGA is a senior associate within the Information Program at the World Resources Institute. Her current work focuses on water resources, global fisheries, and species conservation. She specializes in environmental indicators that measure the condition of ecosystems at the global and regional level, and she is also part of WRI's Global Forest Watch team, coordinating forest monitoring activities with Global Forest Watch partners in Chile. Ms.

Revenga is lead author of the WRI report *Pilot Analysis of Global Ecosystems: Freshwater Systems* (2000) and a contributing author to the WRI's *Pilot Analysis of Global Ecosystems: Coastal Ecosystems* (2001). These two reports assess the condition of freshwater and coastal ecosystems as well as their capacity to continue to provide goods and services that humans depend on. Ms. Revenga is also the lead author of *Watersheds of the World: Ecological Value and Vulnerability* (1998), which is the first analysis of a wide range of global data at the watershed level. Before joining WRI in 1997, she worked as an environmental scientist with Science and Policy Associates, Inc., an environmental consulting firm in Washington, D.C. Her work covered topics in sustainable forestry and climate change.

ROBIN WHITE is a senior associate with the World Resources Institute, an environmental think tank based in Washington, D.C. Her focus at WRI has been on the development of environmental indicators and statistics for use in the *World Resources Report* and in global ecosystems analysis. She was the lead author of the WRI report *Pilot Analysis of Global Ecosystems: Grassland Ecosystems* (2000), which analyzes quantitative information on the condition of the world's grasslands. Her current work focuses on developing an ecosystem goods and services approach to the analysis of the world's drylands. A recent publication regarding this work is WRI's Information Policy Brief, *An Ecosystem Approach to Drylands: Building Support for New Development Policies.* Ms. White completed her Ph.D. in geography at the University of Wisconsin, Madison, with a minor in wildlife ecology. Before joining WRI in 1996, she was a policy analyst with the U.S. Congress, Office of Technology Assessment.

Contributors

MARY KRANE DERR is a freelance writer and editor specializing in environmental, biomedical, and social service issues. Her clients include organizations and journals based in Africa, India, the United States, and Europe. Derr earned a bachelor's degree in biology from Bryn Mawr College and a master's degree in social work from the University of Chicago. Derr's poetry has been nominated for the anthologies Best American Poetry and Best Spiritual Writing and has received honors from the Poetry Center of Chicago and various environmental groups. Currently she is editing *Works of Mercy: Poetic Aftermaths of An Gorta Mor, Ireland's Great Hunger,* an anthology of poetry and short lyrical prose written by descendants of survivors of the mid-nineteenth century Irish potato famine. Derr lives in Chicago with her husband and daughter. She is an honorary member of a Xhosa family living near Cape Town, South Africa, where she was a delegate and presenter at the 1999 Parliament of the World's Religions.

Josh Harkinson is a San Francisco-based freelance journalist. He writes regularly about the environment for newspapers and national magazines. He holds an A.B. in Environmental Science and Policy from Duke University and an M.J. from the UC Berkeley Graduate School of Journalism. Harkinson has traveled in Africa extensively and has written on subjects ranging from gold mining in Guyana to eco-travel in Madagascar.

Phia Steyn is a lecturer in the Department of History at University of the Free State, Bloemfontein, South Africa. A graduate of the University of the Free State (B.A., 1996; M.A., 1998), her Ph.D. thesis is on oil politics and environmental issues in Nigeria and Ecuador. Her work on Africa history and environmental issues has been published in numerous academic journals, including *New Contree, South African Historical Journal, Journal for Contemporary History,* and *Historia.*

Population and Land Use

In terms of population growth, Africa and the Middle East are among the fastest-growing regions in the world. Indeed, Africa's rate of population increase is unequalled on any other continent, including Asia. But many nations of Africa and the Middle East are struggling to provide basic services and economic opportunities for their burgeoning populations. In the Middle East, the presence of the world's largest oil reserves has produced notable standard of living gains in health and education as well as in infrastructure investment. But many people have shared only marginally in this bounty, and the subordinate political and social status of women in Arab nations in the Middle East and North Africa precludes the region from making full use of its total human potential. In addition, expanding populations have made sustainable use of freshwater and other natural resources a major concern throughout the region.

The state of Africa's peoples is far more precarious. Impoverishment is widespread and underpins major shortcomings in public health, education, living standards, and environmental protection. Desperate economic straits have also contributed to the devastating famines and civil wars that have wracked the continent over the past three decades, and poverty (both state and household) is undoubtedly a key factor in the horrific spread of HIV/AIDS across sub-Saharan Africa. Finally, Africa's steadily rising sea of impoverished people is eroding the continent's natural wealth, as fragile land and water resources are repeatedly manipulated and used in damaging ways to stave off hunger and other immediate threats.

People of the Middle East and North Africa

The nations of the Middle East and North Africa (defined here as the African continent's Arabic-speaking Mediterranean states—Morocco, Tunisia, Algeria, Libya, and Egypt) are located in one of the most arid and water-scarce corners of the world. Desert conditions prevail across much of this territory, and marine water resources (such as the Mediterranean Sea and the Arabian Gulf) and freshwater resources (the Tigris-Euphrates) are vital elements of the landscape for human communities and animal and plant biodiversity alike.

Despite the forbidding climatic conditions found in the interior of these countries, this region has a rich human history. Over time, its people have exerted a profound influence over the natural and human history of three neighboring continents—Asia, Africa, and Europe—as well as the world as a whole. Indeed, the Middle East and North Africa constitute "one of the cradles of civilization and of urban culture. Three of the world's major religions originated in the region—Judaism, Christianity, and Islam. Universities existed in [the Middle East and North Africa] long before they did in Europe. In modern times, [its] politics, religion, and economics have been inextricably tied in ways that effect the globe" (Roudi 2001).

Population Trends

For many millennia, human populations in this region hovered at about 30 million. But by the early twentieth century that number had grown to 60 million, and in the second half of the century medical advances (such as antibiotics and immunization measures) and improved sanitation systems made possible a sustained surge in population. The total population of the Middle East and North Africa grew from 100 million in 1950 to around 380 million by 2000, a rate of population growth unmatched anywhere else during that time (United Nations 2001).

Today the region's human populations continue to grow at a steady clip, adding nearly 7 million people every year (ibid.). In mid-2002, the population in the Middle East (also known as Western Asia) stood at 197 million, but forecasters predict that the number could well double by 2050. The mid-2002 population in North Africa, meanwhile, was approximately 180 million. But forecasters say that by 2050, the region's population could break the 300 million mark (Population Reference Bureau 2002). If these forecasts hold true, the total population of the region could reach 700 million by 2050—seven times what it was only a century earlier.

The majority of the region's people now live in urban areas, a marked contrast from only two decades ago, when most people resided in rural settings.

From 1980 to 1998, however, the urban population in the Middle East and North Africa nearly doubled, from 83 to 164 million. This trend has been attributed partly to continued expansion of oil extraction and production and other industrial activities, which are usually located in population centers, and partly to desertification of agricultural lands. Today North Africa ranks as the most urbanized region of the entire continent, with 64 percent of its population living in urban centers. Libya is the most urbanized nation, with 88 percent of its population living in settlements sprinkled along the Mediterranean coastline (UN Centre for Human Settlements 2001).

Financial benefits accruing from its possession of most of the world's known petroleum reserves have enabled the Middle East to weather some aspects of this population surge. For example, several Middle Eastern states have dramatically upgraded their health services sectors in the past half-century. Infant mortality rates in some oil-rich Persian Gulf states such as Kuwait are now roughly comparable to those in the developed world.

Overall, however, there is considerable room for improvement in this regard. In the Middle East, the infant mortality rate of 45 per 1,000 live births ranks below the world average (of 54 deaths per 1,000 live births), but far above the North American (6) and European (8) averages. In North Africa, the infant mortality rate is roughly equivalent to the world average (Population Reference Bureau 2002). As a whole, the infant mortality rate in the Middle East and North Africa remains higher than that of other developing regions, such as Latin America and East Asia (United Nations 2001). Moreover, the region's maternal mortality rate is four times that of East Asia (UN Development Programme 2002).

These comparatively high numbers are an important contributor to current life expectancy levels. Life expectancy in the Middle East and North Africa is sixty-eight and sixty-six years, respectively, which is considerably higher than the regional average of two decades earlier (fifty-nine years). But these numbers are lower than the average for developed countries (seventy-six years) and only slightly higher than the average for all developing nations (sixty-five years) (Population Reference Bureau 2002).

In terms of age demographics, one-third of the population in the Middle East and North Africa is under the age of fifteen, and the number of women of childbearing age (fifteen to forty-nine years) is expected to double in many countries over the next thirty years. In coming years, these trends could well exacerbate unemployment rates, which are already among the highest in the world, and claim governmental resources that would otherwise be devoted to health care and other services for growing populations of the elderly (Roudi 2001). Certainly, the rising number of people of all age groups poses difficult funding allocation questions for governments through-out the region. For example, Egypt's overall fertility rate—the average number of children a woman is forecast to have given current trends—is 3.5. But the country's elderly population—those sixty years and older—is also escalating. In fact, Egypt's elderly population has been projected to grow from 4.3 million in 2000 to 23.7 million by 2050 (United Nations 2001; Roudi 2001).

The region has made significant strides in the education realm. Indeed, secondary school enrollment in the Middle East and North Africa increased from 42 to 64 percent from 1980 to 1997 (World Bank 2001). Nonetheless, stubbornly high levels of illiteracy point to the need for greater government investment in educational programs. In Egypt, for example, the literacy rate among the population age fifteen years and older increased from 40 percent to 50 percent from 1980 to 1995, but the total number of illiterate Egyptians nevertheless grew from 16 million to 19 million. In Morocco, the number of illiterate citizens rose from 8 million in 1980 to 9.5 million in 1995, despite an overall increase in its literacy rate from 29 percent to 44 percent (UN Educational, Cultural and Scientific Organization 1999).

About two-thirds of the illiterate adults in the Arab world are women, which underscores the basic patriarchal nature of the region's societies. Women in Middle Eastern and Northern African states do not enjoy the same rights and privileges as women in the developed world, though the degree of repression varies from country to country. In countries wherein fundamentalist Islamic leaders strongly influence social and political life, women lead extremely circumscribed lives. The Saudi Arabian government, for example, forbids women from driving or voting, forces them to wear veils and robes over their bodies in public, and refuses to acknowledge a woman's right to freely choose her own husband. In Kuwait, the rights to vote or hold public office remain elusive for the country's women. Practices such as forced female circumcision and so-called honor killings, in which male family members murder female family members who bring dishonor to the family (by getting pregnant out of wedlock, for instance, or arousing suspicions that they are guilty of adultery), persist in some Arabic countries. Even in countries like Jordan, where women are free to work, vote, secure an education, and hold political office, some laws place women in a subordinate role. Jordanian law, for instance, makes it impossible for a woman to secure a passport without the permission of a male guardian (a spouse, father, or brother).These limits imposed on women's development are a major factor in the region's overall poor standing in terms of civil and political freedoms. Indeed, according to the United Nations, the region's Arab countries ranked last in civil and political freedoms out of seven world regions (UN Development Programme 2002).

Women's literacy rates in many Arab communities have risen significantly in the past three decades, however, and more equitable rights in such realms as property ownership and voting are slowly being introduced in some countries. In addition, improved access to family planning services has helped to blunt "social attitudes and norms that exclusively stress women's reproductive role" (ibid.). Indeed, most Arab countries support the provision of family planning information and services, and countries such as Algeria, Egypt, Iran, Jordan, and Turkey have adopted explicit policies to lower fertility and encourage smaller families (Roudi 2001).

Nonetheless, many male political, religious, and community leaders in the Arab world continue to resist making women equal partners in charting a path in the new millennium—to the ultimate detriment of their countries. "Women . . . suffer from unequal citizenship and legal entitlements, often evident in voting rights and legal codes," acknowledged one analysis. "The utilization of Arab women's capabilities through political and economic participation remains the lowest in the world in quantitative terms. . . . Qualitatively, women suffer from inequality of opportunity, evident in employment status, wages and gender-based occupational segregation. Society as a whole suffers when a huge proportion of its productive potential is stifled, resulting in lower family incomes and standards of living" (UN Development Programme 2002).

Economic Trends

The Middle East's oil wealth has enabled states to invest considerable sums in various forms of infrastructure, such as schools and highways. But export revenue from the sale of petroleum varies considerably from nation to nation, and all countries—regardless of their level of national income or economic development—are laboring to meet the basic needs of their growing populations. Specific challenges include adequate housing, sanitation, health care, education, and employment, all of which are essential in combating poverty and improving overall standards of living (Roudi 2001).

Another pressing issue in some locales is violence and political unrest, which can evaporate disposable income and devastate businesses. For example, economic activity in the occupied territories has been bludgeoned into a mere trickle by the repeated episodes of violence and terror that have erupted between Palestinians and Israelis. Elsewhere, international sanctions levied against Iraq in the wake of the 1991 Persian Gulf War have devastated the economy. Electric, transportation, water, and sanitation systems are in varying stages of collapse, and personal income has plummeted. Literacy rates and health barometers also declined markedly during the 1990s, though President

Saddam Hussein's role in these trends is a subject of considerable debate. According to a 1999 UNICEF survey, for instance, child mortality in Iraq dropped to pre–Gulf War levels in areas where the United Nations administered food aid. But where Hussein oversaw the administration of this aid, child mortality rates doubled, suggesting that some of the aid was kept or diverted by the ruler (Pollack 2003). In 2003 a U.S.-led military invasion removed Hussein from power, but the socioeconomic future of Iraq remains clouded in uncertainty.

One problem facing much of the region is the continued lack of diversification in economic activity. "Extracting petroleum for export, like the process of extracting other forms of mineral wealth, does not inherently lead to other forms of economic activity in the country in which the extraction takes place," observed one analysis. "The process of producing oil is technologically advanced and capital intensive, requiring a highly skilled—but relatively small—workforce. Therefore, the petroleum sector cannot contribute much to national employment goals in countries with rapidly growing populations" (Center for Strategic and International Studies 2000). This is one major reason why unemployment rates in the region have remained so stubbornly high over the past two decades. In addition, efforts to reduce unemployment are bucking the steady rise in the number of citizens seeking jobs. In Egypt, for example, economists are working feverishly just to avoid losing ground in employment. After all, the economy must create an additional 500,000 new jobs annually just to absorb new entrants into the job market (Roudi 2001).

In recent years, major oil-producing states have repeatedly sought to parlay their fossil fuel wealth into more diverse economies, but gains have been modest. In most of these nations, oil revenue is controlled by the state, which distributes wealth in accordance with its domestic priorities. "Because of their control over oil revenues, governments that are inefficient, corrupt and wasteful are able to remain in power long after they would have been seriously threatened in more normal political circumstances. More important, perhaps, is the fact that the private sectors of these countries become dependent on state guidance, credit, protection and subsidies, and they tend to lose—or fail to develop—the dynamic, innovative, risk-taking and independent spirit required for a healthy nation-building effort" (Center for Strategic and International Studies 2000). This lack of diversification could have devastating consequences for economies throughout the region in the future if environmental and energy security concerns prompt the United States and other major oil markets to satisfy a greater percentage of their energy needs with domestic resources or renewables.

On the other hand, if the region diversifies its economy—and makes genuine commitments to political stability, improved health and education, and equitable economic policies—it will be much better equipped to stem the "brain drain" that is currently afflicting many states. Many Middle Eastern and North African countries feature large, young populations that are well educated and ready for challenging careers. But limited opportunities in this regard have convinced many educated young people to prove themselves in Europe, North America, and other, more economically vibrant, parts of the world (Roudi 2001).

Land Use Trends

Rapid population growth across the region is placing heavy pressure on natural resources, with freshwater scarcity the most obvious threat. Combined, the Middle East and North Africa contain approximately 6.3 percent of the world's population, but the region holds only 1.4 percent of the world's accessible freshwater, making it acutely vulnerable to severe water shortages if present consumption and population trends do not change.

Between 1970 and 2001, for example, population growth in North Africa and the Middle East sliced the available renewable freshwater resources per person by more than half, and forecasts suggest that per capita freshwater availability for the region as a whole will probably decline by 2025 to about 1,000 cubic meters, the internationally recognized threshold for water scarcity. Today, much of the region is already below the international standard, since nearly 80 percent of available freshwater in the region is concentrated in Iran, Iraq, Syria, and Turkey. In countries such as Israel, Jordan, Kuwait, and Saudi Arabia, the annual national average is already below 200 cubic meters per person (Population Action International 2000; Roudi 2001).

Population growth and attendant development pressure are also taking their toll on other natural resources. This is especially true in some smaller states with limited land area. In Israel's case, for instance, environmentalists assert that "much of the country's open space—so vital for leisure and recreation, the preservation of a varied landscape, and assuring the quality-of-life and well-being—has been ceded to dwellings, roads, power lines, and other trappings of development" (Yoav Sagi in Schorsch 2002).

Critics also contend that ecologically insensitive land-use policies are a major factor in the rapid pace of land conversion. In the 1990s, for example, Israel's government guaranteed farmers that sold property for housing developments 30 percent of the development profits. In light of this change, large swaths of farmland that had long enjoyed significant protections from development via the country's municipal planning and construction laws came on the market and were quickly snapped up. In Iraq, meanwhile, it has been estimated that the country has destroyed 85 percent of the Mesopotamian wetlands—the nation's primary source of freshwater—in ill-conceived engineering projects over the last three decades (Lash 2002).

Unsustainable forms of land use and exploitation are also shrinking the size of the region's productive land area. The World Bank has estimated the cumulative impact of land degradation in the Middle East/North Africa region at about U.S.$1.15 billion per year in lost agricultural productivity, and it has estimated that environmental problems account for approximately 14 percent of the region's total health burden. Of this total, about 8 percent is attributable to water supply and sanitation and about 3 percent to urban air pollution (World Bank 2001). As in every other corner of the world, impoverished communities suffer the most from these threats, as low-income housing is often plagued by unreliable or unsafe drinking water, inadequate or nonexistent sanitation systems, and marginal locations that place inhabitants at greater risk from floods, landslides, sandstorms, and other natural disasters. Moreover, high population densities and limited land areas make it economically and politically difficult for some states to add to their protected area networks, which remain modest by international standards.

Finally, the violent conflicts and civil unrest that have bedeviled parts of the Middle East and North Africa and made life a struggle for so many people have also taken a heavy toll on land, water, and air. For example, the protracted conflict between Israel and the Palestinians in the Occupied Palestinian Territories has diminished the quality of "water reserves, soils, forests, and native species, possibly beyond repair," according to the UN Environment Programme. Problems include overgrazing, deforestation, and rising concentrations of toxic waste—all attributable at least in part to restrictions on movement within the territories. But water management looms as the region's biggest environmental problem. "The absence of treatment plants in the occupied territories means wastewater is being dumped on open land. This is squandering freshwater and contaminating aquifers that supply it. Untreated sewage is also ending up in the ocean, polluting Gaza's coastline" (UN Environment Programme 2003).

In the Persian Gulf region, meanwhile, the land is still recovering from the grim ecological legacy of the 1991 Gulf War. The military might of U.S. forces and their allies accounted for some of this environmental damage, but much of the devastation was wrought by Iraqi president Saddam Hussein and his military. As they fled Kuwait, Iraqi forces set fire to approximately 600 oil wells across the country. As much as 6 million barrels of oil a day erupted from the burning wells, submerging lowland habitats and generating horrific levels of

Smoke from oil well fires obscures daylight in Kuwait, 1991. DEPARMENT OF DEFENSE

air pollution. "The oil that did not burn in the fires traveled on the wind in the form of nearly invisible droplets resulting in an oil mist or fog that poisoned trees and grazing sheep, contaminated freshwater supplies, and found refuge in the lungs of people and animals throughout the Gulf. . . . [In addition], the deposition of oil, soot, sulfur, and acid rain on croplands up to 1,200 miles in all directions from the oil fires turned fields untillable and led to food shortages that continue to this day" (Lash 2002). Obeying Hussein's directives, Iraqi forces also intentionally dumped another four million barrels of oil into the Persian Gulf, tarring pristine beaches, killing an estimated 25,000 birds, and diminishing fishery health for the next century (ibid.).

These and other incidents make it clear that political stability and peace are a necessary component of any regional effort to address population challenges and improve policies in such areas as environmental protection, poverty reduction, and human development—and especially raising the status of women. "As the world moves into the twenty-first century, the Arab world is at a crossroads," concluded the UN Development Programme. "While some countries have done well in terms of income and material wealth, human development remains low in many instances. Poverty and deprivation in their many forms remain real in many Arab societies. . . . The fundamental choice is whether the region's trajectory in history will remain characterized by inertia, including the persistence of institutional structures and types of actions that have produced the substantial development challenges it faces, or whether prospects will emerge for an Arab renaissance that will build a prosperous future for all Arabs, especially coming generations" (UN Development Programme 2002).

The People of Sub-Saharan Africa

Much of sub-Saharan Africa is typified by harsh, unforgiving desert and semiarid climatic regimes, though dense tropical forests adorn the continent's midsection. These forests, rich in flora and fauna and natural resources ranging from timber to water, progressively give way to savanna, grasslands, and deserts as one moves southward through the continent. The peoples who live in this part of the continent are remarkably diverse, encompassing thousands of ethnic groups that speak an estimated 1,000 different languages. In many areas, ethnic identity supersedes national or political ties—a contributing factor to the turmoil that has enveloped so many sub-Saharan countries over the years.

The people of sub-Saharan Africa are still emerging from a 500-year period of foreign colonization that ended only in the mid-twentieth century. But legacies of this colonial era abound across the continent, from artificially drawn political boundaries to political, economic, and social systems that were hastily erected in the first months of Africa's postcolonial age. Today, economic and social development across most of the region remains stunted, hampered by political unrest, civil wars, ethnic and racial tensions, and the vagaries of weather and international policy-making. All of these elements are factors in the emergence of a vast population of refugees—nomads uprooted from their ancestral homes by civil strife, environmental disaster, or the threat of starvation. Indeed, the third largest settlement in Sierra Leone today is a refugee camp (UN Centre for Human Settlements 2001), and many other countries have large refugee populations. In some countries, social progress, economic betterment, and environmental awareness are on the upswing. But hunger, disease, and impoverishment remain all too common in every city and many rural communities. As a result, most African nations occupy the lowest rungs of the UN Development Programme's Human Development Index (HDI), which ranks countries around the world on various socioeconomic indicators.

People of Sub-Saharan Africa

As of mid-2002, the human population in sub-Saharan Africa stood at approximately 693 million. Based on current trends, however—such as a fertility rate of 5.6 that is double the world average of 2.8—the regional population is projected to rise by more than 130 percent from current levels by 2050, reaching

Figure 1.1 Position of the Arab Region Compared to Other Regions in the World on Human Development Indicators, 1998

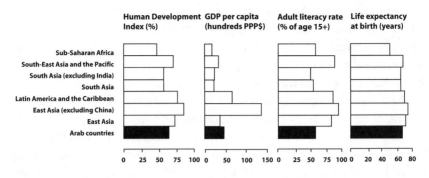

SOURCE: UN Development Programme. 2002. *Arab Human Development Report: Creating Opportunities for Future Generations.* http://www.undp.org/rbas/ahdr/PR2.pdf. Accessed July 22, 2003.

over 1.6 billion. The region's infant mortality rate of 91 per 1,000 live births is nearly double the world average of 54, and the life expectancy at birth for a native of sub-Saharan Africa is forty-nine years. In some countries, life expectancy is shockingly low. In Zambia, for example, life expectancy is only thirty-seven years. By contrast, life expectancy for the developing world as a whole is sixty-five, and in developed countries it is seventy-six. Other nations with horrendous life expectancy rates include Botswana (thirty-nine), Sierra Leone (thirty-nine), and Malawi (thirty-eight) (Population Reference Bureau 2002).

The short life spans that prevail in these nations are directly attributable to impoverishment that has left people exceptionally vulnerable to diseases

Table 1.1 Health Progress and Setbacks in African Countries

Subregion	Life expectancy at birth (years)		Infant mortality (per 1,000 live births)		Under-five mortality rate (per 1,000 live births)	
	1970–1975	1995–2000	1970	1999	1970	1999
Northern Africa (6)	52	66.1	123.8	38.7	190.7	51.2
Western Africa (16–1)	42.7	49.7	161.3	106.3	273.3	174.2
Central Africa (8–1)	43.7	49.1	139.6	103.6	233.3	163.2
Eastern Africa (8–1)	44.7	50.1	133.6	94.7	212	147.9
Southern Africa (11)	47.7	46.2	127.6	91.3	205.1	142.2
IOC (4.1)	—	—	—	—	—	—

SOURCE: UN Environment Programme. 2002. *Africa Environment Outlook: Past, Present and Future Perspectives.* http://www.unep.org/aeo/244.htm. Accessed July 22, 2003.

(such as malaria and HIV/AIDS), malnutrition and famine, and violence. An estimated half of the African population—about 340 million people—live on less than U.S.$1 per day, and fewer than 6 in 10 have access to safe water (Obasi 2002). And while average per capita daily calorie consumption in Africa has increased slightly since 1970, population growth has doubled the total number of undernourished people living in the region (UN Food and Agriculture Organization 2001).

The ability of people engaged in subsistence farming—which remains the dominant economic activity in the region—to lift themselves out of these desperate circumstances is constrained by limited economic opportunities in other sectors, meager educational opportunities, diminished returns from fields that are being exhausted by unsustainable practices, and inequitable land tenure policies dating back to the colonial era that too often place the bulk of premium agricultural land in the hands of a privileged few. In South Africa, for example, black farmers' access to land averages just over 1 hectare per capita, while white farmers have access to 1,570 hectares per capita (Southern Africa Development Community 2000). Land reform efforts have made some progress in this regard in recent years, reducing centralized control and increasing rates of individual and communal ownership, but change has been slow.

The discouraging—and sometimes deadly—conditions that exist in many rural areas has exacerbated migration into cities and towns, most of which are already failing to provide basic services to their existing populations. Africa's rate of urbanization—3.5 percent per year—is the highest in the world. Urbanization is highest in Central Africa and the Western Indian Ocean Islands (each region is 48 percent urban), but it is also accelerating in West Africa (currently 38 percent urban), Southern Africa (36 percent), and Eastern Africa (26 percent). Among individual countries, the highest rates of urbanization in sub-Saharan Africa can be found in Djibouti (83 percent) and Gabon (73 percent), while nations with small urban populations include Rwanda (5 percent), Burundi (8 percent), Ethiopia (15 percent), and Burkina Faso (15 percent) (UN Centre for Human Settlements 2001; Population Reference Bureau 2002).

All told, Africa now holds forty cities with populations of 1 million or more, but overall population growth and migration patterns have been forecast to push that number to seventy by 2015. Lagos, Nigeria, is the largest of these cities, with more than 13 million residents crowding its streets (UN Centre for Human Settlements 2001). So-called unplanned settlements ring all of these urban centers, but living conditions in these areas are generally deplorable. Only a minority of children in these settlements are able to regularly

Table 1.2 HDI Ranking of African Countries in 2000

	African countries by HDI levels		
Subregional grouping	*Low HDI*	*Medium HDI*	*High HDI*
Northern Africa	Sudan (143)	Libya (72) Tunisia (101) Algeria (107) Egypt (119) Morocco (124)	none
Western Africa	Togo (145) Mauritania (147) Nigeria (151) Côte d'Ivoire (154) Senegal (155) Benin (157) Gambia (161) Guinea (162) Mali (165) Guinea Bissau (169) Burkina Faso (172) Niger (173) Sierra Leone (174)	Cape Verde (105) Ghana (129)	none
Central Africa	Democratic Republic of Congo (152) Central African Republic (166) Chad (167)	Gabon (123) Equatorial Guinea (131) Sao Tome and Principe (132) Cameroon (134) Congo (139)	none
Southern Africa	Zambia (153) Tanzania (156) Angola (160) Malawi (163) Mozambique (168)	South Africa (103) Swaziland (112) Namibia (115) Botswana (122) Lesotho (127) Zimbabwe (130)	
Eastern Africa	Djibouti (149) Uganda (158) Eritrea (159) Rwanda (164) Burundi (170) Ethiopia (171)	Kenya (138)	none
Western Indian Ocean States	Madagascar (140)	Seychelles (53) Mauritius (71) Comoros (137)	none

SOURCE: UN Environment Programme. 2002. *Africa Environment Outlook: Past, Present and Future Perspectives.* http://www.unep.org/aeo/023.htm. Accessed July 22, 2003.

Busy street market in Lagos, Nigeria. DANIEL LAINE/CORBIS

attend school or receive basic medical care, and the slums are crowded and bereft of basic sanitation and decent housing. In Congo, for instance, only 14 percent of the country's urban population have access to basic sanitation services. As a result, urban landscapes across the country are heavily scarred by black clouds of pollution from burning refuse and rotting garbage dumps of monstrous size (World Health Organization/UNICEF 2000). Obviously, these conditions provide fertile conditions for the spread of disease.

The poverty that persists across much of sub-Saharan Africa is particularly hard on women, who carry a disproportionate burden of household and caregiver duties in many families. Across much of the continent, ethnic traditions and state laws effectively keep land and control of natural resources in the hands of men. Moreover, women receive discriminatory treatment in such areas as salary, access to credit, and land ownership. It is hoped, however, that land reform efforts will help women gain a more equitable share of the land and resources that remain cornerstones of economic development. "Study after study has shown that there is no effective development strategy in which women do not play a central role," declared UN Secretary General Kofi Annan. "When women are fully involved, the benefits can be seen immediately: families are healthier; they are better fed; their income, savings and reinvestment go up. And what is true of families is true of communities and, eventually, of whole countries" (Annan 2003).

Mourners carry a victim of AIDS to a cemetary for burial. Seventy percent of the world's HIV/AIDS victims live in southern Africa. LOUISE GUBB/CORBIS SABA

The AIDS Epidemic

The HIV/AIDS disease is now a pandemic across sub-Saharan Africa. At the beginning of the twenty-first century, it was estimated that AIDS had already killed nearly 22 million Africans, and that another 34 to 36 million Africans are afflicted with the disease. Most of these people are in sub-Saharan Africa, home to 70 percent of the adults and 80 percent of the children living with HIV around the world (Hunter 2000). In some sub-Saharan nations a quarter or more of the population are believed to be infected, including Zimbabwe (33.7 percent), Swaziland (33.4 percent), Lesotho (31 percent), and Zambia (25 percent). South Africa, meanwhile, has the largest total number of people now living with HIV/AIDS in the world—4.2 million (Joint UN Programme on HIV/AIDS 2000a, 2002).

In these and other developing African nations, resources available to fight the epidemic, from hospitals and clinics to funds for education and treatment programs, are completely inadequate to stem the rising tide of misery and death. In fact, allocations for all health-related investments—including AIDS—by the central governments of many African countries in the 1990s have been less than 10 percent, and in some cases as little as 5 percent, of the governmental budget (UN Children's Fund 2001).

Given the financial straits in which African countries find themselves—a situation exacerbated by crushing debt burdens to Western creditors and all-too-commonplace evidence of official corruption and inefficiency—reliance on international assistance in addressing the growing HIV/AIDS problem is considerable. Some level of assistance has been proffered thus far; for example, Western pharmaceutical companies have sent relatively inexpensive versions of AIDS regimens to affected countries in Africa at a fraction of their cost. "Realistically, however, national health budgets and individuals' incomes in the poorest countries—where the majority of the population lives on less than US\$1 a day—still will not allow widespread access to the drugs, even at relatively low prices" (Deame 2001a). During the late 1990s, meanwhile, overseas donations to combat the disease amounted to less than U.S.\$100 million annually. Still, there are signs that global recognition of the size of the African HIV/AIDS crisis is on the rise. In early 2003, for example, U.S. president George H. Bush pledged \$15 billion for programs to fight HIV/AIDS on the continent.

But experts caution that money alone will not curb the devastation. Deep cultural stigmas about AIDS and sexually transmitted diseases that exist throughout much of the continent make open discussion and education programs difficult to implement. In addition, while some countries—including Uganda, Senegal, and Zambia—appear to be making measurable progress in reducing HIV infections through aggressive education and communication campaigns, other governments have shown little political will to address the issue and its causes squarely (Akukwe and Foote 2002). Others simply divert money from health care to other priorities. The government of Chad, for example, recently diverted \$4.5 million of an initial \$25 million World Bank grant meant for health care to weapons for its war against secessionists (Robinson 2002).

In the meantime, the disease is taking its toll in myriad ways. The loss of adult family members to AIDS is pushing many poverty-stricken households into even more desperate conditions, as is the cost of treatment. In places like Tanzania and Zaire, for instance, studies indicate that households with AIDS often lose a year's annual income paying for AIDS treatment and funeral costs (Hunter 2000). The pandemic is also stretching limited health care resources so thin that quality of care in other areas (such as treatment of malnutrition and newborn care) has suffered. In many countries, AIDS patients are simply crowding other non–HIV infected populations out of existing health care systems. The AIDS scourge is also depriving countries of people who fulfill valuable roles in the community. Scores of dedicated nurses, schoolteachers, park administrators and rangers, and business people have been lost to the epidemic (Deame 2001b).

Table 1.3 Percentage Change in Indebtedness by African Countries

Country	Total external debt U.S.$ (million) 1985–1987	Total external debt U.S.$ (million) 1995–1997	Percentage change
Angola	4,035	10,739	166
Benin	1,012	1,611	59.2
Botswana	438	626	42.9
Burkina Faso	659	1,286	95.1
Burundi	598	1,117	86.8
Cameroon	4,003	9,394	135
Central African Republic	474	921	94.3
Chad	275	975	255
Congo	3,625	5,439	50
Congo, D. R.	7,373	12,799	73.6
Côte d'Ivoire	11,562	18,010	55.8
Equatorial Guinea	162	286	76.5
Eritrea	–	52	–
Ethiopia	6,234	10,155	62.9
Gabon	1,923	4,318	125
Gambia	281	437	55.5
Ghana	2,779	5,992	116
Guinea	1,767	3,334	88.7
Guinea Bissau	390	918	135
Kenya	4,841	6,922	43
Lesotho	211	669	217
Liberia	1,461	2,091	43.1
Madagascar	3,073	4,191	36.4
Malawi	1,182	2,253	90.6
Mali	1,749	2,970	69.8
Mauritania	1,740	2,405	38.2
Mozambique	3,496	5,833	66.8
Namibia	—	—	—
Niger	1,411	1,567	11
Nigeria	23,392	31,318	33.9
Rwanda	474	1,061	124
Senegal	3,275	3,725	13.7
Sierra Leone	870	1,169	34.4
Somalia	1,816	2,628	44.7
South Africa	—	25,543	—
Sudan	9,945	16,967	70.6
Tanzania	6,506	7,345	12.9
Togo	1,078	1,427	32.4
Uganda	1,522	3,652	140
Zambia	5,655	6,933	22.7
Zimbabwe	2,631	5,006	90.3

SOURCE: UN Environment Programme. 2002. *Africa Environment Outlook: Past, Present and Future Perspectives.* http://www.unep.org/aeo/014.htm. Accessed July 22, 2003.

Finally, the HIV/AIDs crisis in Africa is generating a staggering number of orphans. In some sub-Saharan African countries, AIDS will kill an estimated one-fifth to one-third of all adults over the next seven to ten years unless massive efforts are made to provide antiretroviral treatments to those affected (Hunter 2000). This grim harvest will create a generation of orphans (most of them HIV negative) that are at high risk of receiving inadequate health care, uncertain access to food and water, minimal education, and meager emotional sustenance during their formative years.

Some of these orphans will be taken in by extended family, but AIDS has shredded many communities so completely that some orphans have no relatives left. Orphanages, meanwhile, are impractical in many countries, either because governments cannot afford to build and maintain them or because orphans who leave their home village must give up claims to their parents' land under some existing land tenure systems (Deame 2001a).

At the close of the twentieth century, the United Nations estimated that 90 percent of the 13 million orphans (children who lost their mother or both parents by age fifteen) created by the HIV/AIDS crisis were located in sub-Saharan Africa (Joint UN Programme on HIV/AIDS 2000b). Moreover, it is believed that in eight of the countries in the region, more than 20 percent of the children under fifteen are orphans of AIDS or other causes of death (Hunter 2000). These numbing statistics make it clear that "AIDS is changing the social landscape in the most affected countries. It's creating an unprecedented set of child welfare problems" (Hunter and Williamson 2000).

The Economy of Sub-Saharan Africa

Because of the continued economic struggles of most sub-Saharan African nations, Africa is the only continent where poverty levels are widely expected to rise in the twenty-first century. Indeed, the continent's economic pulse is incredibly faint, given the size, population, and natural resources at its disposal; in 1999, only 1.1 percent of global gross domestic product (GDP) was contributed by Africa (World Bank 2000).

In essence, these nations are trapped in a vicious circle. They desperately need economic growth to improve the standard of living of citizens and fund health, education, and environmental programs. But political instability and civil strife, corruption, conflicts with neighboring states, massive debt, and health and education shortfalls make it difficult to attract investment and generate new business activity. This, in turn, keeps the populace impoverished and bereft of the educational opportunities that are essential in encouraging business growth and entrepreneurial activity.

The impact of these factors is further intensified by economic policies—many of them in place since the end of the colonial era—that have been ineffective. Policies and administrative developments that have drawn particularly strong criticism include wage and price controls, subsidies of basic goods and commodities, bloated and inefficient civil service sectors, fixed currency exchange rates (which promote overvaluation of currencies), high tax rates, and various other disincentives to foreign investment. In response to these conditions, the World Bank and the International Monetary Fund have both insisted that recipient countries adhere to a variety of economic adjustments, ranging from reforms to the civil service to removal of domestic subsidies. But countries have chafed at these conditions, citing the social and economic problems—spikes in unemployment, inflationary prices for bread and other basic commodities, and the like—that have resulted (UN Environment Programme 2002).

Natural disasters can also be especially punishing to the economies and infrastructure of countries in sub-Saharan Africa, since they do not have the financial resources to withstand such blows and rebuild quickly. In Mozambique, for instance, an ambitious development program was launched after its long civil war ended in 1992. But flooding in 2000 from tropical cyclone Eline destroyed nearly one-fifth of the country's only highway and extensive sections of railway linking Mozambique to neighboring Zimbabwe. The damage from these floods was eventually estimated at nearly 12 percent of the country's total gross national product. Struggling African economies simply do not have the capacity to shrug off this sort of punishment (Cornford 2001; Obasi 2002).

A number of sub-Saharan African nations do have considerable natural wealth in the form of minerals, timber, water (in the tropical rain forest countries), and fisheries. But countries have thus far failed to harvest these resources in an ecologically sensitive manner, or maximize the benefits—employment, skills development, income, and foreign exchange—that could accrue from sustainable exploitation. Whatever the reasons for this wastefulness—poorly conceived policies, official malfeasance, or political maneuvering—it is not in the long-term best interests of the countries, their citizenry, or their flora and fauna (Ascher 2000).

Some examples of the squandering of natural resources are truly startling. Angola has considerable oil and diamond riches, but exports of these materials are funding a bloody civil war that has killed 500,000 people and turned another 2 million into refugees (the government relies on oil revenue, while rebels buy arms from diamond sales). With most of the money generated within the country devoted to warfare, cities such as Luanda, the Angolan capital, are falling apart. Reliable access to water and electricity is rare in many regions of the country, and life expectancy and literacy rates are plummeting. Moreover, the Angolan population is increasingly reliant on food imports, as tillable areas are now infested with millions of land mines. "Angola is perhaps the perfect example of a nation plunged into despair because of the tragic manipulation of its natural resource endowments," summarized one institution, which adds that the basic dynamics in the country are unlikely to change any time soon: "To date, the civil war has not disrupted oil exploration and production activities, in that the major fields are offshore Cabinda and isolated from the fighting. . . . [But] if the past is prologue, then little of the coming oil wealth will pass to the benefit of the Angolan people. Rather, the hard currency earned from oil exports more likely will be employed in support of efforts by the MPLA [government] to subdue all opposition, and that, in turn, can only translate into further devastation" (Center for Strategic and International Studies 2000).

Land Use in Sub-Saharan Africa

The impoverishment that characterizes much of sub-Saharan Africa is closely intertwined with the environmental degradation and resource depletion that threatens the region. Indeed, the deteriorating state of Africa's land and water is both a cause and a consequence of the poverty and malnutrition that dog many regions. Increasingly frantic efforts to provide for the social, educational, health, and economic needs of populations are producing a host of destructive land-use practices, from large-scale deforestation (in tropical forests) to conversion of wetlands and other ecologically vital habitats for agriculture and human settlement. Contamination of air and water from pollution-intensive industries and energy sectors is also on the rise.

Across Africa, great swaths of land have suffered some level of degradation in the last half-century. More than 20 percent of the continent's vegetated land is now classified as degraded, with most of these areas categorized as moderately to severely degraded. In addition, the number of countries that are regularly vulnerable to water shortages is steadily rising. In 2000 about 300 million Africans lived in a water-scarce environment. By 2025 the number of countries in Africa experiencing water stress will rise to eighteen, affecting an estimated 600 million people. Rampant desertification from unsustainable farming practices, overgrazing, and deforestation—all made more acute by population growth—is perhaps the most frightening phenomenon on the continent, for it is driving Africa's growing incapacity to feed its people. Conditions of desertification are now manifest on an estimated 13 million square kilometers—43 percent of the continent's total land area—on which 270 million people live (Obasi 2002; UN Environment Programme 2002).

In rural areas, poverty and population growth are putting increasing pressure on mammals and birds, including some threatened species. Farmers desperate to boost their yields are increasingly venturing onto marginal land that nonetheless serves as important natural habitat, and poaching and trade of wild animal meat—commonly known as bushmeat—are surging. In urban areas, meanwhile, unplanned settlements are extending tendrils deeper into the countryside, enveloping natural habitat, wreaking changes on populations of numerous species of flora and fauna, and altering the ecological functions of marshlands, streams, savannas and woodlands. Erosion-prone hillsides, natural drainage waterways, and flood-prone areas have been particularly vulnerable to the impacts of urbanization (UN Environment Programme 2002). Nearly two decades ago, an observer of the continent's problems commented that "unless sudden massive strides are made in development, burgeoning populations in Africa translate to human misery—and disaster for the land, which will be called upon to be increasingly more productive" (Rosenblum and Williamson 1987). Today, these words ring even more true, and sustainable development initiatives are even more urgently needed.

Finally, the African landscape and the millions of people who depend on it for their survival will undoubtedly experience a whole new set of challenges if predicted manifestations of global climate change come to pass. According to the Intergovernmental Panel on Climate Change, global warming will likely exacerbate drought and famine conditions in many areas of Africa, which has far fewer institutional and economic resources to prepare for such eventualities than most other areas of the world. The IPCC predicts that more severe water shortages will result in even greater declines in the production of food and heighten stress on already oversubscribed natural resources at a time when brisk population growth is predicted. "The greatest impact [of climate change] will continue to be felt by the poor, who have the most limited access to water resources," stated the IPCC. "Water shortages will damage inland fisheries, through drought and habitat destruction. Ocean warming also will modify ocean currents, with possible impacts on coastal marine fisheries. Temperature rises will spread diseases such as malaria. Droughts and floods, where sanitary conditions are poor, will increase the frequency of epidemics. . . . Increased temperatures of coastal waters could aggravate cholera epidemics in coastal areas. In addition, sea-level rise, coastal erosion, saltwater intrusion, and flooding will have significant impacts on African communities and economies" (Intergovernmental Panel on Climate Change 2001).

Building Better Lives in Sub-Saharan Africa

"The African calamity is easy to ignore," commented one analysis of the economic, social, and environmental forces buffeting the continent's peoples. "Sustained attention is hard and depressing work. Instead, the world acts when an 'emergency' is declared; it relaxes when the danger is pronounced past. . . . The human suffering, like the factors that cause it, is seen from a distance in abstraction. Up close, the calamity is anything but abstract" (Rosenblum and Williamson 1987).

Certainly, charting a new course for sub-Saharan Africa will be a daunting task, requiring skillful navigation through deadly reefs and shoals of poverty, malnutrition, institutional apathy, and environmental degradation. At this juncture, "the existing infrastructure and capacity to ensure sustainable development are inadequate in Africa," acknowledged one expert. "Virtually no country, region, or sector is well placed to meet the unique needs of integrating across many different scientific disciplines or of fostering the interaction necessary for making progress in sustainability science. An appropriate enabling environment must be developed. Generating adequate scientific capacity and institutional support is particularly urgent" (Obasi 2002).

But experts believe that the states of sub-Saharan Africa—working in concert with international agencies and nongovernmental organizations—can make measurable progress in improving living standards and husbanding natural resources for future generations if they can make gains in the following areas:

IMPROVED REGIONAL COOPERATION
Each country in sub-Saharan Africa confronts unique challenges in its efforts to improve public health and education, reduce civil unrest and strife, ensure basic human rights, and preserve natural resources for future generations. But the obstacles to these goals—poverty, malnutrition, poor governmental performance, unsustainable exploitation of land, air, and water—are ultimately the same for all of these nations, and they can be tackled more effectively if individual countries embrace a spirit of cooperation with neighboring states. Toward this end, in 2001 the Organization of African Unity Summit passed a New African Initiative that specifically declares a goal of eradicating poverty across the continent and placing all African countries, individually and collectively, "on a path of sustainable growth and development." The New African Initiative identifies a host of sectoral priorities and expands on existing initiatives to address all manner of challenges, including environmental protection, smart urban development, economic diversification, health care, energy and transportation, and global warming.

NEW ECONOMIC INITIATIVE AND REFORMS

Many African countries are slowly dismantling their inefficient state-centered economies, and the International Food Policy Research Institute (IFPRI) asserts that "for the first time since independence, continent-wide and subregional perspectives on development solutions are gaining strength and visibility. This shift toward greater ownership of the development agenda opens the door for more countries to benefit from greater economic integration and to capture spillover benefits from the exchange of technology and information." The IFPRI cites the New Partnership for Africa's Development (NEPAD) as a particularly promising initiative among African leaders that "concretely reflects the continent-wide commitment to ownership of future development priorities," and urges the introduction of other regional trading arrangements along the lines of the Southern Africa Development Corporation (SADC) free trade area, which was formally launched in 2000 (International Food Policy Research Institute 2002).

IMPROVED MANAGEMENT OF LAND AND WATER

The states of sub-Saharan Africa have to dramatically improve their management of limited—and dwindling—freshwater resources if they are to have any hope of making inroads in such issues as health and nutrition, sustainable economic development, and biodiversity protection. "Water quality and quality assessments should go hand in hand, and the planning and management of water resources must take into account the likelihood of floods and droughts to occur. . . . [In many countries,] competition for water for agricultural, domestic, and industrial purposes is clearly evident," and conflicts over water between households, farms, and industries will only intensify, barring major changes in use and distribution.

In the realm of land care, nations such as Ethiopia, Kenya, Malawi, Senegal, and South Africa have all proven that restoration programs can improve soil fertility and productivity. But sustainable agricultural practices must be introduced on a large scale to reduce rates of desertification and nutrient depletion that are shrinking the region's area of arable land. "Desertification, poverty, development pressures and climatic factors interact in a complex manner to influence food security. It is, therefore, essential that desertification be tackled within a development framework, and in a participatory manner. The approach must combine: political and legal reform; economic and social development strategies; land tenure reform; international partnerships; capacity building; and financial sustainability" (UN Environment Programme 2002).

Similarly, new models of sustainability and long-term planning have to be integrated into the operating blueprints of African cities, where exploding populations are currently overwhelming extremely limited municipal resources. Priorities in this regard should include: investment in family planning programs to curb the rate of population growth; greater emphasis on environmentally and socially sensitive development projects; abandonment of failed economic policies that stifle growth; basic reforms to institutional structures (both public and private) to reduce corruption and increase operating efficiencies; reduction of debt burden (such as through "debt for nature" swaps pioneered by Kenya); reconfiguration of budget priorities at the federal level to better address basic education and health care needs; and implementation of rural development programs that can slow the migration of villagers to urban centers (ibid.; UN Centre for Human Settlements 2001).

Sources:

Abrams, Len. 2002. "Drought and Famine in Southern Africa: A Review of the Crisis." *The Water Page*, Available at http://www.thewaterpage.com/drought_crisis_2002.htm (accessed February 2003).

Akukwe, Chinua, and Melvin Foote. 2002. "HIV/AIDS in Africa: The Death March Continues." *Africa News Service*, (July 14).

Al-Mughni, Haya. 2001. "All Roads Lead to the Franchise." *UNESCO Courier* (March).

Albert, Jeff, Magnus Bernhardsson, and Roger Kenna, eds. 1998. *Transformations of Middle Eastern Natural Environments: Legacies and Lessons.* New Haven, CT: Yale University Press.

Annan, Kofi A. 2003. "To Save Africa, We Must Save Africa's Women." *Africa News Service* (February 10).

Ascher, William. 2000. "Understanding Why Governments in Developing Countries Waste Natural Resources." *Environment* 42 (March).

Bloch, Marianne, Josephine Beoku-Betts, and B. Robert Tabachnick, eds. 1998. *Women and Education in Sub-Saharan Africa: Power, Opportunities and Constraints.* Boulder, CO: Rienner.

Brooks, Geraldine. 1995. *Nine Parts of Desire: The Hidden World of Islamic Women.* New York: Doubleday.

Center for Strategic and International Studies. 2000. *The Geopolitics of Energy into the 21st Century.* Washington, DC: CSIS.

Cornford, S. G. 2001. "Human and Economic Impacts of Weather Events in 2000." *World Meteorological Organization Bulletin* 50 (October).

Deame, Laura. 2001a. "A Generation of Orphans: Another Challenge for AIDS-Ravaged Countries." In *Earthtrends: The Environmental Information Portal,* World Resources Institute (May). Available at http://earthtrends.wri.org (accessed February 2003).

———. 2001b. "Global AIDS Toll Bleak." In *Earthtrends: The Environmental Information Portal,* World Resources Institute (June). Available at http://earthtrends.wri.org (accessed February 2003).

Goodwin, Jane. 1994. *Price of Honor: Muslim Women Lift the Veil of Silence on the Islamic World.* Boston: Little, Brown.

Hunter, Susan S. 2000. *Reshaping Societies: HIV/AIDS and Social Change.* Glen Falls, NY: Hunter Run.

Hunter, Susan S., and John Williamson. 2000. *Children on the Brink 2000.* Washington, DC: USAID.

Intergovernmental Panel on Climate Change. 2001. *Climate Change 2001: Mitigation, Impacts, Adaptation, and Vulnerability: Summaries for Policymakers.* Geneva: IPCC.

International Food Policy Research Institute. 1997. *The World Food Situation: Recent Developments, Emerging Trends, and Long-Term Prospects.* Washington, DC: IFPRI.

———. 2002. *Ending Hunger in Africa: Only the Small Farmer Can Do It.* Washington, DC: IFPRI.

James, Valentine Udoh. 1995. *Women and Sustainability in Africa.* New York: Praeger.

Joint UN Programme on HIV/AIDS. 2000a. "AIDS Epidemic Update." Geneva: UN-AIDS.

———. 2000b. *Report on the Global HIV/AIDS Epidemic.* Geneva: UNAIDS.

———. 2002. *Report on the Global HIV/AIDS Epidemic 2002.* Geneva: UNAIDS.

Lash, Jonathan. 2002. "The Environment: Another Casualty of War?" World Resources Institute (December). Available at http://jlash.wri.org (accessed February 2003).

Obasi, Godwin O. P. 2002. "Embracing Sustainability Science: The Challenges for Africa." *Environment* 44 (May).

Pollack, Kenneth M. 2003. *The Threatening Storm: The Case for Invading Iraq.* New York: Random House.

Population Action International. 2000. *People in the Balance: Population and Natural Resources at the Turn of the Millennium.* Washington, DC: PAI.

Population Reference Bureau. 2002. "2002 World Population Data Sheet." Washington, DC: PRB.

Robinson, Simon. 2002. "Scarred: War, Bad Government, and AIDS are Feeding a Deadly Drought across Southern Africa." *Time International* (August 5).

Rosenblum, Mort, and Doug Williamson. 1987. *Squandering Eden: Africa at the Edge.* New York: Harcourt Brace Jovanovich.

Roudi, Farzaneh (Nazy). 2001. "Population Trends and Challenges in the Middle East and North Africa." Population Reference Bureau (December). Available at http://www.prb.org (accessed February 2003).

Schorsch, Jonathan. 2002. "Israel: A Land of Asphalt and Cement?" *Tikkun* 17 (January/February).

Southern Africa Development Community. 2000. *SADC Human Development Report: Challenges and Opportunities for Regional Integration.* Harare, Zimbabwe: SAPES.

Southern Africa Development Community, World Conservation Union-IUCN, Southern African Research and Documentation Centre, and Zambezi River Authority. 2000. *State of the Environment in the Zambezi Basin 2000.* Harare, Zimbabwe: SADC, IUCN, SARDC, ZRA.

Stillwaggon, Eileen. 2002. "HIV/AIDS in Africa: Fertile Terrain." *Journal of Development Studies* 38 (August).

Stolberg, S. 2001. "Africa's AIDS War." *New York Times,* March 10.

United Nations. 2001. *World Population Prospects: The 2000 Revision.* New York: UN.

UN Centre for Human Settlements. 2001. *The State of the World's Cities 2001.* Nairobi: UNCHS.

UN Children's Fund-UNICEF. 2001. *State of the World's Children 2001.* New York: UNICEF.

UN Development Programme. 2002. *Arab Human Development Report 2002.* Available at http://www.undp.org/rbas/ahdr (accessed February 2003).

UN Educational, Cultural and Scientific Organization. 1999. *UNESCO Statistical Yearbook 1999.* Paris: UNESCO.

UN Environment Programme. 2002. *Africa Environment Outlook.* Nairobi, Kenya: UNEP.

———. 2003. *Desk Study on the Environment in the Occupied Palestinian Territories.* Available at http://www.unep.org/GoverningBodies/GC22/Information_documents.asp (accessed February 2003).

UN Food and Agriculture Organization. 2001. *The State of Food and Agriculture 2001.* Rome: FAO.

World Bank. 1999. *Confronting AIDS: Public Priorities in a Global Epidemic.* New York: Oxford University Press.

———. 2000. *World Development Indicators 2000.* Washington, DC: World Bank.

———. 2001. *Middle East and North Africa Region Environmental Strategy Update.* Washington, DC: World Bank.

World Health Organization and UN Children's Fund (UNICEF). 2000. *Global Water Supply and Sanitation Assessment 2000.* Geneva and New York: WHO and UNICEF.

Biodiversity
—JOSH HARKINSON

Humans and wildlife have coexisted in Africa and the Middle East longer than anywhere else on earth. But although a few extinctions at the hands of humans occurred long ago in the region, most damage to the continent's biodiversity has come with rapid human population growth—and attendant development—ushered in during the twentieth century. Today, the outlook for biodiversity protection is bleak, particularly in many parts of Africa, where high rates of population growth, severe economic poverty, high levels of corruption, and political unrest are the norm. Indeed, these realities are an explosive combination for flora and fauna and the ecological systems upon which they depend, for African states struggling to provide basic necessities for their peoples are unlikely to make major investments in conservation, including protected area systems. In the neighboring Middle East, meanwhile, steadily rising rates of natural resource consumption and land conversion have also produced rising numbers of threatened and endangered animals and plants. Indeed, most nations of the Middle East have consciously opted for ambitious development projects and oil and gas exploitation schemes at the expense of natural ecosystems.

Biological Wealth in Africa and the Middle East

Biological diversity, or biodiversity, is generally defined as the variety of life forms found in a given region (whether an ecosystem, nation, continent, etc.), but also including genetic diversity within species. Indicators of overall levels of biodiversity include species richness (number of species), species diversity (types of species), and levels of endemism (species found nowhere else).

As is the case around the world, only a small fraction of the total number of species inhabiting Africa and the Middle East have been identified (although cataloging of mammal and bird species has been more extensive than that of other taxonomic groups such as plants, invertebrates, and insects). Moreover, known species of flora and fauna in Africa and the Middle East are less likely to have been subject to extensive scientific monitoring and research than those in North America, Europe, Australia, or parts of Asia and Latin America. Despite this paucity of biodiversity research, however, scientists have recorded more than 50,000 plant species, 1,000 mammal species, and 1,500 bird species on the African continent. But a substantial number of these species experienced steep declines in range and population over the past century. During that time, it has been estimated that Africa lost a total of 126 recorded animal species to extinction (either total extinction or extinction from the wild), and more than 2,000 other animal species are now known to be threatened. In addition, 123 recorded African plants are now extinct, and more than 17,700 others are threatened (UN Environment Programme 2002). Many species in the Middle East are under imminent threat of extermination as well. For instance, Turkey has 78 animal species that are classified as vulnerable, endangered, or critically endangered, and significant numbers of vulnerable or endangered animals have also been reported by nations with comparatively modest numbers of animal species, including Israel (41 species), Iran (55), and Saudi Arabia (27) (World Conservation Union 2002).

Africa's biological resources are the "backbone of the African economy as well as the life-support system for most of Africa's people" (UN Environment Programme 2002). For example, plant and animal resources are harvested extensively for food; house, boat, and cart construction materials; and as raw materials for manufactured goods. Other resources, such as trees and crops, are traded commercially. In addition, some species have provided genetic material used in the creation of drought- and disease-resistant crops and other agricultural products. The Middle East, though best known for its economic reliance on its large oil reserves, is also heavily dependent on its biological capital for a great deal of its economic activity.

For example, biodiversity is an important component in the tourist industries of Africa and the Middle East, which are vital to the economic fortunes of some states. In 1995, for example, tourism provided 24 percent of Tanzania's foreign exchange earnings (Kaiza-Boshe 1998). In Kenya, tourism was the second largest currency earner, behind agriculture, throughout the 1990s (Mugabe 1998). The preservation of habitats, charismatic species, and other elements of regional ecosystems is thus an economic as well as ethical priority in some places.

A less fickle source of revenue than tourism—medicines from Africa's plants—has benefited people for generations. Drugs from medicinal plants meet the health needs of 80 percent of the population in Ethiopia, where some 800 beneficial species are found (Tedla 1998). In South Africa, according to FAO reports, about 20,000 tons of medicinal plants are traded each year, with a value of about $60 million. Although medicinal uses provide an incentive for biodiversity conservation, they can also become a threat without proper management. Some 27 million South Africans use medicinal plants, and their popularity is so great that many species have become extinct outside protected areas.

One great hope is that Western pharmaceutical companies will eventually contribute money for the protection of African biodiversity, supplying key funds for management and conservation. Such assistance would address historical inequities in the disposition of revenue generated by medications developed from African plant species. In 1958, for example, indigenous medicine men in Madagascar pointed foreign researchers to the Madagascar rosy periwinkle. From this small pink flower, scientists derived two alkaloids effective against Hodgkin's disease and childhood leukemia. Global sales of these drugs earn foreign companies an estimated $100 million each year, but neither Madagascar nor the medicine men who led the researchers to the plant ever saw a dime of the profits. In the future, African countries may begin demanding financial agreements with pharmaceutical companies before allowing them to comb forests for valuable materials.

Threats to Biodiversity in Africa and the Middle East

The biodiversity contained within each specific habitat type in Africa and the Middle East faces unique threats to the habitats upon which they depend for survival. These include the logging industry in tropical forests; desertification and overgrazing from livestock in the savanna and semiarid regions; dams and irrigation projects in wetlands; and regional impacts from agriculture, urbanization, and hunting. Additional threats underpin some of these factors and apply across the African continent, such as poverty, high rates of human population growth; the harvest of species for fuelwood and bushmeat; the effects of political instability and war; international trade in endangered species; invasion by alien (non-native) species; and lack of enforcement of existing laws to protect biodiversity.

Growing Human Populations

Since the 1990s, Africa's population has grown at the fastest rate in the world, and in 2001 the continent's estimated human population reached 840 million. Despite recent signs that population growth may be tapering off because of improved family planning and the AIDS scourge, population growth of as much as 120 percent has been forecast for the next half-century (Population Reference Bureau 2002). Population trends point to higher demand for and consumption of natural resources in the Middle East as well. In 2001 the human population in West Asia—composed primarily of Middle Eastern countries—reached 197 million, and estimates of 100 percent population growth for the region in the next fifty years have been made (ibid.).

These expanding populations continue to exert pressure on protected and unprotected natural habitats alike, especially in Africa. In many African countries, parks and protected areas offer only nominally greater levels of security for species and their habitats than unprotected tracts of land or marine areas, because of funding shortfalls, institutional indifference, and resource consumption pressure. Moreover, neither the size nor number of parks, game reserves, and conservation areas in Africa is likely to increase significantly in the near term. In most African countries, the desperate human need for land for agriculture and other purposes has prevented governments from setting aside more protected areas.

Growing human populations consume so much land in Africa because the economy is undeveloped. More than 70 percent of Africans live in rural areas, and the majority of them still depend on subsistence agriculture for survival (Perrings 2000). Thus when populations expand, so does the demand for arable land. Additionally, because few markets for domestic goods exist in Africa, a majority of Africans rely heavily on wild plants and animals for sustenance. For example, a government study in Cameroon found that half of local incomes earned by rural households were generated by subsistence extraction of food from local forests. Population growth magnifies this drain on biodiversity as direct demands on species and their habitats increase beyond sustainable limits.

The most common use of wild species for resources is the exploitation of wooded areas for fuel. Many Africans must rely on wood for fuel because electricity is unaffordable or unavailable. Use of fuelwood is most destructive near dense urban areas, where people collect wood faster than forests can regenerate, keeping them in a permanently degraded state. In some areas, communities have begun to replant forests for later use as fuel, and more efficient woodstoves have been promoted in many places. Nevertheless, overuse of fuelwood remains a widespread problem for biodiversity, linked to economic underdevelopment.

Another factor contributing to declines in the populations of some animals is Africa's increasing demand for bushmeat. According to the UN Food and Agriculture Organization, bushmeat supplies 84 percent of protein for many Nigerian communities, and accounts for 20 to 25 percent of meat consumption in rural Tanzania (Osei 1997). People often kill or buy wild game because they cannot afford meat from domestic livestock. In Zimbabwe, for example, the wildlife trade monitoring group TRAFFIC has reported that bushmeat is 75 percent cheaper than pasture-raised alternatives. Moreover, some urban Africans continue to eat wild game because they feel it ties them to cultural traditions. Long-standing beliefs concerning the alleged benefits of consuming certain animals add to the pressure. In some parts of Africa, for example, eating monkey is believed to make one clever, while gorilla meat is thought to confer strength.

In the past, rural communities often harvested game meat sustainably, but many modern hunters are commercial poachers toting automatic weapons, and the construction of logging roads has provided them with the means to reach previously inaccessible areas. Beginning in the 1980s, commercial hunters overran ancestral hunting ranges across Africa's equatorial belt, including Guinea, Liberia, Ivory Coast, Ghana, Cameroon, Gabon, the Democratic Republic of Congo, and Congo. Timber companies contributed to the trend; instead of importing food, they found it cheaper to ask workers in the field to buy bushmeat. This new demand encouraged traditional hunters to abandon their snares for guns in order to supply the commercial market. TRAFFIC reports that in Tanzania, more than a third of traders rely on bushmeat as their sole source of income (Verrengia 2001).

The effects of this trade on populations of some wild animals has been devastating, prompting organizations such as the UN Food and Agriculture Organization to warn of a "bushmeat crisis" in equatorial Africa. "The forests of tropical Africa are rich in primate species, which are particularly vulnerable to overexploitation because they breed slowly and often have small populations. About 15 primate species are believed to be threatened by the bushmeat trade. The number of chimpanzees in Africa is believed to have declined by 85 percent during the twentieth century. Other species threatened by the bushmeat trade include the forest elephant, the water chevrotain, six duiker species, the leopard, and the golden cat" (UN Food and Agriculture Organization 2001b). Indeed, although satellite images of forest tracts in parts of West and Central Africa show an intact canopy cover, field surveys reveal that many of these forest patches have been "emptied out" of animals.

International Trade in Endangered Species

As a menace to overall biodiversity, the international trade in endangered species pales in comparison to problems like habitat loss. Nevertheless, several species have declined precipitously at the hands of poachers who sell them as pets or export their horns, tusks, and hides to buyers in the Western world and

Asia. For example, demand for rhino horn was the major factor in the collapse of black rhino populations from 65,000 in 1970 to 2,600 in 1998 (Hutton and Dickson 2000). People in Yemen and Oman wear daggers with rhino horn handles as status symbols, and traditional doctors in China use powdered horn to treat delirium, convulsions, and fever. Other species affected by international trade include elephants (killed for their ivory), crocodiles (for skin), and parrots, Egyptian tortoises, and other species popular in the international pet trade.

In 1973 some 150 nations signed the Convention on International Trade in Endangered Species of Wild Fauna and Flora, known as CITES, for the express purpose of halting or limiting this type of commerce. The treaty lists threatened species under three separate categories, or Appendices, that permit varying degrees of trade. Appendix I regulates species like the black rhino, prohibiting trade except for noncommercial purposes such as scientific research. Commercial trade in Appendix II species, such as the Nile crocodile, is allowed but highly regulated. Nations can place species unlisted on Appendix I and II on Appendix III if they wish to limit trade across their own borders. For example, Ghana lists 116 species and subspecies of bird on Appendix III, more than any other country. CITES also allows countries to make "reservations" on certain species to circumvent bans on trade. Only two African countries hold reservations, Namibia with the cheetah and Malawi with the African elephant.

The effect of CITES on Africa's endangered species has been mixed. A ban on the trade in rhino horns, imposed in 1977, has done little to stem declines in the species. After the ban went into effect, black market prices for rhino horn increased tenfold in some countries. The lucrative market stimulated more poaching in countries lacking the financial resources to enforce the ban. For this reason, some African countries have asked for permission to trade in rhino horns and then use the profits to fund additional conservation measures. Opponents of a limited, legalized trade say it would just supply cover to

Table 2.1 Threatened Species: Country Totals by Taxonomic Group (2002 Red List)

Africa	Mammals	Birds	Reptiles	Amphibia	Fishes	Molluscs	Other Inverts	Plants	Total
North Africa									
Algeria	13	6	2	0	1	0	11	2	35
Egypt	13	7	6	0	0	0	1	2	29
Libyan Arab Jamahiriya	8	1	3	0	0	0	0	1	13
Morocco	16	9	2	0	1	0	7	2	37
Tunisia	11	5	3	0	0	0	5	0	24
Western Sahara	3	0	0	0	0	0	0	0	3

(continues)

(continued)

Table 2.1 Threatened Species: Country Totals by Taxonomic Group (2002 Red List)

	Mammals	Birds	Reptiles	Amphibia	Fishes	Molluscs	Other Inverts	Plants	Total
Sub-Saharan Africa									
Angola	19	15	4	0	0	5	1	19	53
Benin	8	2	1	0	0	0	0	11	22
Botswana	6	7	0	0	0	0	0	0	13
Burkina Faso	7	2	1	0	0	0	0	2	12
Burundi	6	7	0	0	0	0	3	2	17
Cameroon	40	15	1	1	27	1	3	155	243
Cape Verde	3	2	0	0	1	0	0	2	8
Central African Republic	14	3	1	0	0	0	0	10	28
Chad	17	5	1	0	0	1	0	2	26
Comoros	2	9	2	0	1	0	4	5	23
Congo	15	3	1	0	1	1	0	33	54
Congo, The Democratic Republic of the	40	28	2	0	1	41	4	55	171
Côte d'Ivoire	19	12	2	1	0	1	0	101	136
Djibouti	4	5	0	0	1	0	0	2	12
Equatorial Guinea	16	5	2	1	0	0	2	23	49
Eritrea	12	7	6	0	0	0	0	3	28
Ethiopia	35	16	1	0	0	3	1	22	78
Gabon	15	5	1	0	1	0	1	71	94
Gambia	3	2	1	0	1	0	0	3	10
Ghana	14	8	2	0	0	0	0	115	139
Guinea	12	10	1	1	0	0	3	21	48
Guinea-Bissau	3	0	1	0	1	0	1	4	10
Kenya	51	24	5	0	18	12	3	98	211
Lesotho	3	7	0	0	1	0	1	1	13
Liberia	17	11	2	0	0	1	1	46	78
Madagascar	50	27	18	2	14	24	8	162	305
Malawi	8	11	0	0	0	8	0	13	40
Mali	13	4	1	0	1	0	0	6	25
Mauritania	10	2	2	0	0	0	0	0	14
Mauritius	3	9	4	0	1	27	5	87	136
Mayotte	0	3	2	0	0	0	1	0	6
Mozambique	14	16	5	0	4	6	1	36	82
Namibia	15	11	3	1	3	1	0	5	39
Niger	11	3	0	0	0	0	1	2	17
Nigeria	27	9	2	0	2	0	1	119	160
Réunion	3	5	2	0	1	14	2	14	41
Rwanda	9	9	0	0	0	0	2	3	23
Saint Helena	1	13	1	0	7	0	2	9	33
Sao Tome and Principe	3	9	1	0	0	1	1	27	42
Senegal	12	4	6	0	1	0	0	7	30
Seychelles	4	10	4	4	0	1	2	43	68
Sierra Leone	12	10	3	0	0	0	4	43	72
Somalia	19	10	2	0	3	1	0	17	52
South Africa	42	28	19	9	29	10	102	45	284
Sudan	23	6	2	0	0	0	1	17	49
Swaziland	4	5	0	0	0	0	0	3	12
Tanzania, United Republic of	42	33	5	0	17	41	6	235	379
Togo	9	0	2	0	0	0	0	9	20
Uganda	20	13	0	0	27	7	3	33	103
Zambia	11	11	0	0	0	4	2	8	36
Zimbabwe	11	10	0	0	0	0	2	14	37

SOURCE: IUCN. 2002. *2002 IUCN Red List of Threatened Species.* http://www.redlist.org. Accessed August 14, 2002.

additional black market exports, and they express doubt that revenues would be used for conservation purposes. They point to the example of the African elephant, which declined for years before a ban on the ivory trade led to modest increases in some populations.

In a few cases, a sustainable trade in wildlife flourishes through ranching projects. Ostrich farms turn a profit in South Africa and Zimbabwe, where some birds have been used to restock wild areas. Also successful are Nile crocodile ranches, which provide a financial incentive to conserve wild populations. Crocodile ranches take eggs from the wild, raise them, and then rerelease a portion of adults. Although the Nile crocodile is listed under CITES, several countries with sustainable ranching programs have received permission to export the species (CITES 2003).

Alien Species

Alien invasive species constitute another major threat to biodiversity in Africa and the Middle East, particularly in aquatic ecosystems. These exotic plants and animals, introduced into regional ecosystems either accidentally or intentionally, have in many cases thoroughly disrupted fragile ecological systems. Freed from the threat of predators or other factors that limited their population growth in their native habitat, these species can assume a dominant role in foreign habitats, crowding out native species in the process. African island states in the Western Indian Ocean, as well as mainland countries in the continent's southern and eastern reaches, have found themselves in a particularly grim struggle with this problem. For example, Ugandan, Tanzanian, and Kenyan fishermen are all grappling with the repercussions of the introduction of the Nile perch and Nile tilapia to the waters of Lake Victoria in the 1960s. Shortly after their introduction (which was executed intentionally as a way of stimulating regional fishery productivity after years of overfishing and eutrophication in the lake), the Nile perch accounted for only 1 percent of the lake's total fish catch. By 2000, however, it accounted for about 80 percent of Lake Victoria's annual fish harvest, and its introduction is believed to be the primary culprit in the loss of more than 200 endemic species from the lake (Ministry of Finance Planning and Economic Development 2000).

Although the introduced fish devastated the lake's biodiversity, they did not destroy the commercial fishery. In fact, total fish production and its economic value rose considerably as a result of the perch introduction. Unfortunately, local communities that had depended on the native fish for decades did not benefit from the success of the Nile perch fishery, primarily because Nile perch and tilapia are caught with gear that local fishermen cannot afford. And since most of the Nile perch and tilapia are shipped out of the region, the local

The Ivory Trade

Game warden with Kenya Wildlife Services (KWS) sorts through impounded elephant tusks at KWS headquarters in Nairobi, February 25, 2003. Kenya wildlife authorities have expressed concern that the easing of a ban on ivory trade could lead to a rise in poaching elephants. REUTERS NEWMEDIA INC./CORBIS

Ivory has been traded for millennia. Romans, ancient Egyptians, and other Arab peoples all traded ivory internationally on a very large scale. Indeed, Romans may have eliminated elephants north of the Sahara. In more recent times, international ivory trade was driven by European colonizers who swept through the African continent shooting elephants, buying tusks from villagers for a pittance, or simply confiscating them. Exported to Europe, tusks became billiard balls, piano keys, hair combs, and—in an echo of their past function—false teeth. By the peak of the trade in the late nineteenth

century, the original elephant population of some 10 million had probably declined by more than half. The trade did not stop, however, and nearly 100 years later it spiked again, fueled by jeeps, automatic weapons, and a void of effective regulation. Efforts to regulate the ivory trade began in 1976 under the Convention on International Trade in Endangered Species of Wild Fauna and Flora (CITES), but to little avail. Elephant numbers plummeted from about 1.3 million in 1979 to about 630,000 ten years later. At that point, CITES members banned the ivory trade completely, slowing

(continues)

elephant declines but creating new problems (Tesi 2000).

Conservationists continue to clash on the merits of the ivory ban. Publicity from the ban ended demand for ivory in Europe and America, reducing prices and diminishing poaching. In that respect it had the opposite effect from the trade ban on rhino horns, which did little to reduce demand and increased prices. Yet in the principal market for rhino horns—Asia—demand for ivory also remains strong, and large amounts are still smuggled there. In fact, the ivory ban alone has not halted declines in elephant populations, which by 1998 were down to between 300,000 and 500,000 (Hutton and Dickson 2000). After all, the ban does not address other threats to elephants, such as habitat loss and hunting for meat. Several African countries argue that the elephant could follow in the footsteps of the highly endangered rhino if the ivory trade is not managed to provide economic incentives for conservation.

The irony of the elephant conservation dilemma is that in several countries, elephant populations are growing too large for their limited ranges. In Kenya and Zimbabwe, for example, elephants increasingly encroach on areas inhabited by peasant farmers, who sometimes kill them as a threat to crops and safety. In South Africa elephants are confined in fenced areas, and growing populations have begun to degrade rangelands. At CITES meetings throughout the 1990s, Southern African countries asked for permission to trade in elephant hides and tusks, arguing that the money could

be used for conservation efforts. CITES denied the requests because most countries fear that a legal trade in one country could provide cover for ivory laundering from other nations.

At the 1997 CITES meeting in Harare, Zimbabwe, delegates authorized a compromise that may eventually lead to other changes in CITES policy. Botswana, Namibia, and Zimbabwe were allowed to move the African elephant from CITES Appendix I to Appendix II. However, the only ivory sale permitted was a one-time auction of legally acquired reserves to Japan, which occurred in 1999 for $5 million.

After the 1997 sale, there were increases in ivory poaching and smuggling in some areas. This fueled the arguments of Kenya and India, which led a coalition of governments that successfully opposed efforts to reverse the ban in 1999 and 2000. In 2002 a number of countries, including Botswana, Namibia, South Africa, and Zimbabwe, successfully introduced resolutions at the CITES Conference of the Parties for a one-off sale of African elephant ivory stockpiles. In the future, the issue of a resumed trade in ivory is not likely to go away, as nations seek to balance economic development and the preservation of key living resources.

Sources:

Hutton, Jon, and Barnabas Dickson. 2000. *Endangered Species, Threatened Convention*. London: Earthscan.
Tesi, Moses K. 2000. *The Environment and Development in Africa*. Lanham, MD: Lexington.

availability of fish for consumption has declined, exacerbating protein malnu-trition among the people of the lake basin.

In addition, the fish introductions impacted the ecosystems around the lake as well. Native fish species used to be air dried for local consumption, but the Nile perch is oilier and requires firewood. This has increased pressure on the area's limited forests, increasing siltation and eutrophication, which, in turn, has further unbalanced the precarious lake ecosystem.

In many other parts of Africa, invasive species such as water hyacinth are choking waterways and lakes, while prickly pear and other invasive plant species are consuming large quantities of extremely limited freshwater.

Some countries are addressing these problems in innovative ways, however. For example, South Africa launched a Working for Water (WFW) project, an initiative that aims to combine poverty alleviation with water conservation by hiring rural poor to remove prickly pear and other invasive plant species that are sucking up disproportionate shares of the country's water. In Kruger National Park, one of South Africa's premier parks, the program is clearing about 120 hectares (nearly 300 acres) daily, and by mid-2002 it had cleared about 40,000 hectares (98,840 acres) of the park of prickly pear and thirty other alien plant species.

War and Civil Unrest

Without political stability, attempts to conserve biodiversity often fail. Wars create a vacuum of regulation, and armies and refugees can destroy the best-managed wild areas in only months. The militarization of Africa began in the 1960s, when colonies fought for independence from European control. In many cases, these conflicts had relatively little impact on biodiversity. Revolutionaries in Zimbabwe, for example, held a taboo against killing cer-tain species of large mammals, and armies were often fed by locals who of-fered domestic livestock (Chenje and Johnson 1994). In the three decades that followed, however, a buildup of arms and animosities fueled highly de-structive civil wars. In Chad, Sudan, Sierra Leone, Algeria, Ethiopia, Somalia, Mozambique, Angola, and the Democratic Republic of Congo, biodiversity has suffered dramatically on account of the effects of war. In the Middle East, meanwhile, the Persian Gulf War wreaked havoc on habitat in some areas of Kuwait and Iraq. In a few cases, however, wildlife has actually benefited from conflict. In Liberia, for instance, flight from dangerous areas left large sec-tions of the countryside uninhabited and unhunted.

In many countries wracked by civil war, national and rebel armies alike have been known to loot natural resources to feed themselves and pay for weapons. In Angola, for instance, revolutionary UNITA forces have admitted to buying arms from South Africa with ivory and teak. By 1992 an estimated

90 percent of Angola's large mammals were gone, many of them killed for meat (ibid.). In Mozambique, the policy of eating wild animals was overt. Government troops took over a large part of Zinave National Park in the 1980s, evicted the park staff, and set up meat processing centers. According to the WWF, herds of buffalo on the Zambezi delta declined from 55,000 in 1979 to 1,000 in 1996 as a direct result of these sanctioned activities. Hippo and waterbuck populations also plummeted.

The effects of war on biodiversity can continue after the fighting ends as well. In 1994 about 850,000 refugees fled the genocidal war in Rwanda and set up camps in Virunga National Park in the Democratic Republic of Congo. Up to 40,000 people entered the park every day, removing between 410 and 770 tons of forest products and threatening the highly endangered mountain gorilla. The DRC then became embroiled in its own civil war, leaving it with few resources to devote to maintaining its park system (Pierce 1994).

Corruption and Lack of Enforcement Capacity

Many African countries have laws prohibiting poaching and destruction of habitat, but they lack the financial means or institutional will to enforce them. In Tanzania, for example, an estimated 60 percent of the value of the wildlife to the economy derives from illegal uses (Kaiza-Boshe 1998). In Cameroon, illegal logging was the chief factor in the loss of about one-tenth of the nation's forests between 1980 and 1995. In part this problem derives from a lack of enforcement capacity. A 2000 study found that each of the five most heavily forested provinces of Cameroon had only one vehicle at the disposal of forest officials (Tesi 2000). Enforcement staff in many African countries also suffer from inadequate training, overwork, and low pay. This situation increases the temptation for rangers to comply when poachers and timber companies offer bribes. Corruption is even prevalent in many "legal" resource extraction operations. For example, in Cameroon logging concessions are not always awarded to the highest bidder as stipulated by law, and concessions are sometimes granted inside formally protected areas.

Habitat Types and Biodiversity

Africa and the Middle East feature extreme variations in terms of climate and physical geography. This variety has contributed greatly to the region's high levels of species diversity and endemism. For example, the humid tropical forests of equatorial Africa support an estimated 1.5 million species, making them among the richest ecosystems on the planet (World Conservation Monitoring Center 2000). But the deserts of the Middle East and Africa—though terribly inhospitable to many species—also nurture a unique array of plant and animal species that have adapted to the arid environment in which they live.

Savannah grasses are either high grasses or short grasses with groups of palms or single baobab trees scattered throughout. PHOTODISC, INC.

Savanna

The term "savanna" describes grassy ecosystems, ranging from environments with no trees at all to woodlands with bushes and a nearly continuous canopy of branches overhead. The savanna ecosystems cover more than half of Africa, extending from Senegal on the east coast to Somalia in the west, and down the eastern side of the continent to the country of South Africa. The savanna contains about 6,500 endemic plant species, the majority of them in the southern region surrounding Zambia (World Conservation Union 1990). As human populations have grown, savanna vegetation has changed significantly because of cattle grazing, agricultural expansion, and wood harvesting. In Kenya, where the human population has tripled since the early 1960s, 5 percent of the trees and shrubs and 8 percent of the herbaceous plant species are endangered. (Mugabe 1998). Some trees, like the muninga of the miombo woodlands, have been overharvested, even in forest reserves. Valued for furniture, the muninga cannot be produced efficiently in plantations because of its slow growth rate (McClanahan 1996).

During the dry season, which lasts four to eight months in the savanna, wild herds of hoofed mammals, or ungulates, often migrate in search of forage and water. The largest of these migrations span hundreds of kilometers between the Mara woodlands in Kenya and the vast Serengeti grasslands to the south in Tanzania. The herds include several hundred thousand zebra and

Biodiversity and Local Control

A debate throughout Africa centers on how much control over wildlife to give to local peoples. Indigenous communities have not always managed wildlife well. In some villages, cultural beliefs even contribute to the eradication of rare species. In Madagascar, for example, many villagers believe that if an endangered aye-aye crosses their path, they must kill it to prevent the death of someone in the community. In North Africa, locals in some places kill the endangered Barbary hyena because they believe it robs corpses from graves. In other cases, however, local beliefs have helped conserve species that have been eradicated elsewhere. For example, forest habitats of the black and white colobus and the mona monkey are protected around two villages in the Brong Ahafo region of Ghana (in otherwise degraded regional environments) because locals consider the monkeys to be the sons of gods who protect the two communities (Osei 1997).

Regardless of local beliefs, however, experience in Africa shows that usurping local control of resources often leads to problems. In Kenya, for example, where the state owns all wildlife, a ban on hunting instituted in 1977 has done little to prevent declines. Between 1977 and 1998, at least 40 percent of range animals disappeared, according to government surveys. Ranchers contend that they should be allowed to own the wildlife on their lands, so that they will

have a financial incentive to conserve it. But the Kenyan Wildlife Service (KWS) has argued that such efforts will never succeed on a large enough scale to save wildlife, and it maintains that the best solution is to fence in protected areas and combat poaching. However, many ranches are now allowed to harvest or crop wildlife on their land under a KWS permit system.

This strategy contrasts sharply with wildlife management policies in Zimbabwe, where the government has encouraged the participation of the local population. In 1986, Zimbabwe established the Communal Area Management Program for Indigenous Resources, known as CAMPFIRE. The program gives local districts the responsibility of protecting game, setting hunting quotas, drawing up contracts with tour companies and safari-hunting operators, and paying dividends from profits to locals. Widely implemented, the program paid significant profits in some areas. In the northern Zimbabwe village of Masoka, for example, each family received $450 in wildlife sales to a safari company in a single year—a substantial income. Due to the success of the program, privately and communally held land accounted for 50 percent of all habitat allocated for wildlife by the early 1990s (Chenje and Johnson 1994), and it remains a popular program in the early twenty-first century.

Provisions to give local communities more control over forest

(continues)

management and resources are being developed and implemented in Lesotho and Tanzania, and in less pronounced ways in emerging developments in Malawi, Namibia, Uganda, and Madagascar. In some cases, plans have gone back to the drawing board for lack of sufficient local input. The major challenge for these programs, however, will be implementation. Changing land tenure requires massive institutional and governance structures, organization, financial resources, and significant political will in the face of vested private and government interests that benefit from the existing system.

Sources:

Chenje, Munyaradzi, and Phyllis Johnson, eds. 1994. *State of the Environment in Southern Africa*. Harare, Zimbabwe: South African Research and Documentation Centre, IUCN-World Conservation Union, and Southern African Development Community.

Osei, William Yaw. 1997. "Human-Environmental Impacts: Forest Degradation and Desertification." In Samuel Aryeetey-Attoh, ed., *The Geography of Sub-Saharan Africa*. Upper Saddle River, NJ: Prentice Hall.

Perrings, Charles. 2000. *The Economics of Biodiversity Conservation in Sub-Saharan Africa*. Northampton, MA: Edward Elgar.

more than a million wildebeest. Numbers of wildebeest more than quadrupled between 1961 and 1996 because of beneficial changes in the ecology of Serengeti ecosystems. In other cases, however, domestic livestock have caused declines in wild species by transmitting disease. In Mali and Senegal, for example, cattle pass bovine plague to Africa's largest antelope species, the endangered western giant eland. More significantly, large numbers of wild ungulates were slaughtered in Southern Africa as part of ill-advised tsetse fly control programs in the 1960s and 1970s.

Humans and their livestock have also caused declines in wild ungulates by impeding their access to key water sources during the dry season. For example, the endangered Grevy's zebra experiences water scarcity inside protected areas as the result of agricultural diversions from streams. But it also faces water shortages outside park boundaries, where pastoralists (shepherds, herdsmen, and others engaged in animal husbandry) rely on streams and lakes to sustain their herds. Between 1977 and 1988, the Grevy's zebra population in Kenya decreased by 70 percent, while the number of sheep and goats increased by 54 percent. Pastoralists monopolize the valuable forage within range of water holes and also scare the Grevy's zebra away, forcing it to drink at night, when it faces increased risk of predation. Formerly depleted by demand for its hide's thin, elegant stripes, this rarest living species of zebra could go extinct within fifty years unless these problems are meaningfully addressed (Reading and Miller 2000).

As pastoralists exclude wild species, the savanna's carnivores have increasingly resorted to preying on domestic livestock. Predators such as the lion, cheetah, and the brown hyena have subsequently suffered declines from pastoralists who kill them to protect their herds. Small predators, like the African golden cat that roams West African woodlands and the black-footed cats of Zimbabwe, are believed to have maintained stable populations because they are more elusive and require less food and habitat (James 1993).

Forests

Tropical rain forest in Central and West Africa contains fewer species than the rain forests of Asia and Latin America. Nonetheless, the forests of equatorial Africa contain an estimated 1.5 million species, more than half of the total biodiversity in sub-Saharan Africa. The Democratic Republic of the Congo (DRC) contains more than 60 percent of this forest area, making it the epicenter of the region's biodiversity and endemism. Indeed, more than 1,100 species of birds and 400 species of mammals are found within the DRC forests. These forests also harbor an estimated 11,000 plant species, 30 percent of which are endemic to the country (UN Food and Agriculture Organization 2001a; Bryant et al. 1997).

Historically, logging pressure in the forests of equatorial Africa has been limited. In the DRC, for example, exploitation of the old-growth forests of the Congo Basin has been hampered by limited transportation infrastructure. Commercially attractive tree species also tend to be widely scattered in the forest, which relieves pressure from industrial logging operations (UN Food and Agriculture Organization 2001a). Nonetheless, degradation and outright clearing of natural forests is believed to be escalating in the DRC, as well as in other forest-rich countries such as Cameroon, Gabon, and the Congo.

The situation is much worse to the west, in the rain forest countries of Ghana, Cote d'Ivoire, Liberia, Sierra Leone, and Guinea. The World Resources Institute (WRI) estimates that between 1980 and 1990, forested areas in West Africa diminished at 2.1 percent per year, the highest rate in the world. By 1992 the World Conservation Union (IUCN) had reported that less than 20 percent of original forest cover remained in most West African countries. Logging is the primary factor in this habitat loss. Logging activity not only destroys forest habitat but often constitutes only the first stage in a process of complete deforestation. The logging roads, which often must span long distances to reach tree species of commercial value, later become conduits for poachers and farmers. In fact, 90 percent of slash-and-burn agriculture by immigrant farmers is concentrated in recently logged areas (Martin 1991). This unsustainable rate of exploitation has imperiled many species. Cote d'Ivoire, for example, has eighty species of flora and fauna that are classified as vulnerable, endangered, or critically endangered, while Ghana has ninety-three species at risk (World Conservation Union 2002). Specific mammals at risk include the Jentink's duiker, the Liberian mongoose, and the pygmy hippopotamus.

Overall, forest loss and degradation in Africa has resulted in the designation of several continental forest areas as internationally recognized "biodiversity hotspots"—areas where species diversity and endemism are particularly high, and where there is an extraordinary threat of loss of species or habitat (Conservation International 2002). One of these hotspots is the Guinean Forest, a ribbon of fragmented forest that trails down the coast of Western Africa through eleven countries, from Guinea to Cameroon. This region has the highest mammalian diversity of all the world's twenty-five hotspots (551 species out of the 1,150 mammalian species in the African region) and contains 2,250 plant species, 90 bird species, 45 mammal species, and 46 reptile species found nowhere else (Conservation International 2002; Mittermeier et al. 2000). Another African forest hotspot is the Eastern Arc Mountain Forests of Eastern Africa, where ecosystems that have evolved in isolation over millennia have produced high levels of plant and animal endemism (Lovett 1998). Other African forests that are seen as potential hotspots in the twenty-first century include the miombo woodlands of interior southern Africa, the dwindling forests of Madagascar, and the forests of the Albertine Rift in East Africa (Mittermeier et al. 1999).

Mountain Zones

Africa has extensive montane (mountain) zones, widely scattered across Ethiopia, Kenya, Tanzania, Uganda, Rwanda, and the Guineo-Congolean forest. Because of the isolation of these areas, unique species have evolved in each of them, leading to greater overall biodiversity in the region. In sum, about 4,000 plant species inhabit the afromontane zone, of which 80 percent are endemic (World Conservation Union 1990). Further, about 10 percent of these species exist on no more than one mountain (McClanahan 1996). As a result, even small portions of montane areas are biologically important, and many species in the region are endangered or threatened where humans have reduced their habitat.

Montane regions have come under intense exploitation because they often contain fertile volcanic soils ideal for farming. Dense human populations drawn to these regions also cut remaining forest for timber. On Mount Kenya, for example, the endemic Meru oak has been harvested to near extinction. In some cases, montane forests cannot grow back because human

The Mountain Gorillas of East-Central Africa

The mountain gorillas of east-central Africa are among the rarest large mammals on earth. Their plight first came to international attention in the 1980s, thanks in part to the pioneering research of American scientist Dian Fossey. Fossey's experiences among the mountain gorillas helped dispel popular conceptions of the animals as violent and aggressive, which had been fueled by King Kong and other media creations. Her research instead showed these great apes to be shy, easygoing herbivores that live in tight-knit family groups. Fossey's murder in 1985 and the subsequent feature film about her life, *Gorillas in the Mist*, created worldwide concern for the conservation of mountain gorillas.

Mountain gorillas *(Gorilla gorilla beringeih)* are the most critically endangered of the three gorilla subspecies that exist in Africa. Between 650 and 700 mountain gorillas remain in the tropical rain forests of east-central Africa. Around half of these animals live in the Virunga mountains, a chain of extinct volcanoes that straddles the borders of Rwanda, Uganda, and the Democratic Republic of Congo (DRC). Although the Virunga region is ecologically homogenous, it is separated into three protected areas—Volcano National Park in Rwanda, Mgahinga Gorilla National Park in Uganda, and Virunga National Park in the DRC—with a total area of around 300 square kilometers. The Virunga population of mountain gorillas often wanders among the parks and crosses international borders. The remaining 350 mountain gorillas live in the 330-square-kilometer Bwindi Impenetrable Forest National Park in southwest Uganda.

Like many other species in Africa, mountain gorillas face threats resulting from destruction of their tropical rain forest habitat. Logging has limited their range to a few isolated patches of protected forest. In addition, the parks that support the gorillas are increasingly encroached upon to meet the subsistence needs of the region's large and growing human population. Rwanda, for instance, has a population of 8 million people crammed into a land area of only 10,000 square miles, making it one of Africa's most densely populated and poorest countries. In Uganda, settlers and farmers have established villages and planted fields of potatoes right up to the entrance of Virunga National Park.

Another significant threat to the continued survival of mountain gorillas comes from human warfare. All three countries that provide habitat for the subspecies have suffered under brutal dictatorships, political unrest, or ethnic strife. Perhaps the best-known example was the 1994 genocide campaign that took the lives of an estimated 1 million people in Rwanda. The tiny country has long been torn by ethnic conflict between the Tutsis—which make up only 15 percent of the population but constitute most of the ruling class—and the Hutus—which account for 84

(continues)

percent of the population and are mostly poor farmers. A series of violent clashes escalated into a civil war in 1994, when a Hutu-led guerilla group called Interahamwe launched a genocide campaign against Rwanda's Tutsi population.

Throughout this period, the country's parks and preserves served as staging areas and combat zones for the warring factions. In fact, rebels laid thousands of land mines in Volcano National Park. Tens of thousands of refugees moved through the park to escape the violence, poaching wildlife for food and felling trees for shelter and fuelwood. An estimated eighteen mountain gorillas were killed as a direct result of the conflict, which also destroyed the infrastructure of the Volcano National Park and took the lives of many members of its staff.

The political instability and violence that have besieged the Virunga region may also threaten the mountain gorillas by reducing tourism. Some experts view ecotourism as a potential route toward ensuring the gorillas' survival. Before 1994, for example, tourism provided the third-largest source of foreign currency in the Rwandan economy, after coffee and tea exports. It continues to be an important source of revenue for Uganda, particularly through gorilla-watching tours in the Bwindi Impenetrable Forest. Westerners pay as much as $3,000 per person to view mountain gorillas in the wild. Some of these funds subsidize the Ugandan Wildlife Authority budget, while 20 percent goes to aid farmers in the areas surrounding Bwindi. In this way, tourism has helped to create jobs for local people, reduce poaching, and encourage conservation of gorilla habitat.

It remains to be seen whether ecotourism can take hold amid the conflict surrounding the Virunga region. In March 1999, a group of Rwandan Interahamwe rebels who had been living in exile in the DRC crossed into Uganda and captured a group of Westerners on a gorilla-watching tour. The rebels murdered eight of the tourists before releasing the other hostages. This event attracted a great deal of international attention and created security concerns that frightened away untold numbers of potential visitors. Nevertheless, by 2001 Rwandan authorities insisted that Volcano National Park had been cleared of mines and was patrolled by well-armed rangers, making it safe for tourists as well as for mountain gorillas.

Some experts believe that efforts to protect the mountain gorilla can contribute to regional stability in east-central Africa by giving people economic incentives to work together despite their political or ethnic differences. Officials from the three countries that border the Virunga region began meeting in 1997 to define shared goals for gorilla conservation. Several nongovernmental organizations—such as the International Gorilla Conservation Program and the Gorilla Mountain Project—have also sponsored projects that span international borders.

(continues)

In Rwanda, researchers have begun collaborating with American scientists to create detailed digital maps of mountain gorilla habitat. Data from satellites and aircraft are combined with ground surveys to construct these maps of forest resources, which can show different species of plants as well as the destruction of habitat. The scientists involved hope that the mapping project will draw attention to Rwandan conservation efforts and help restore the confidence of tourists.

Sources:

Glanz, James. 2000. "Tracking Gorillas and Rebuilding a Country." *New York Times,* April 11.

Matloff, Judith. 1996. "Above Rwanda's Madding Crowd, Mountain Gorillas Reign." *Christian Science Monitor,* December 10.

Snell, Marilyn Berlin. 2001. "Gorillas in the Crossfire." *Sierra* (November–December).

Stanford, Craig B. 1999. "Gorilla Warfare." *The Sciences* (July).

Stoddard, Ed. 2002. "Hope for Africa's Last Mountain Gorillas." *World Environment News.* Available at http://www.planetark.org/avantgo/dailynewsstory.cfm?newsid=14419 (accessed on February 6, 2002).

Weber, Bill, and Amy Vedder. 2002. *In the Kingdom of Gorillas: The Quest to Save Rwanda's Mountain Gorillas.* London: Aurum Press.

development has trapped migratory herbivores in the highlands. Grazing animals confined in Kenya's Aberdare Mountains, for example, have converted logged areas to savanna. This situation further imperils species dependent on the forest for survival.

As in other montane regions, animals are normally sparse in the afromontane zone, where elevations average above 2,000 meters. Several exceptions exist in the vast Ethiopian highlands, which make up more than half the high-altitude terrain in Africa. Endangered mammals endemic to this region include the walia ibex, the mountain nyala, and the Ethiopian wolf. Fewer than 500 of these wolves remain, confined primarily to Bale National Park (Reading 2000). Many nearby residents lack land, and some of them have killed wolves in retribution against the government for restricting their use of parklands.

The birds of Africa's highlands rival all other montane fauna for their diversity. According to Bird-Life International, the highest concentrations of threatened bird species in Africa inhabit the afromontane region, particularly in the Albertine Rift Mountains of the DRC, Rwanda, Burundi, Uganda, and the Eastern Arc mountains of Kenya and Tanzania. Tanzania has the most threatened bird species of any country—and the most threatened species of flora and fauna overall in Africa, with 379 (World Conservation Union 2002).

Mediterranean

The North African countries bordering the Mediterranean Sea have climates akin to those of Spain and Italy and contain species with affinities to both Africa and Europe. The vegetation resembles the chaparral of California. Pine, oak, and cedar forest probably covered much of the wetter areas at one time, but now only a few fragments remain. Because of this loss of habitat, more than 10 percent of mammal species in North Africa are threatened, the highest percentage for any region of similar size on the continent (UN Secretariat of the Convention on Biological Diversity 2001).

Humans have significantly altered the North African environment for thousands of years, and the loss of major species from the region began a long time ago. Elephants probably inhabited North Africa in Roman times, and lions began disappearing in many areas in the 1700s. The last Barbary lions (the North African subspecies) were seen in parts of Morocco in the 1920s. Another predator, the Atlas bear, was the only bear species in Africa before it disappeared from Morocco at the end of the nineteenth century.

The predators that still survive in North Africa are among the most threatened in the region. For example, the Barbary hyena subspecies numbers only 400 to 500 and is limited to the Morocco-Algeria border, a fraction of its previous range. The largest leopard subspecies also inhabits this area, reduced significantly from its original range over most of northwest Africa. Threatened ungulates include several species of gazelle confined to parts of Morocco and Algeria. At the western border of Tunisia, the Barbary deer is still vulnerable, although it has benefited from protected areas for the last few decades.

Cape Region

The climate and scrubby vegetation of the Cape environment resembles that of North Africa, being dominated by "fynbos," a hard-leafed, evergreen, and fire-prone shrubland that thrives on rocky or sandy nutrient-poor soils. Within the fynbos, the vegetation includes grasses and heath as well as various types and sizes of shrubs. However, the biodiversity found in this, the southernmost tip of the continent, is much richer than that present in the north. Although only about half the size of North Dakota, the Cape contains a higher concentration of plant species than any other area of similar size in Africa, and perhaps the world. About 8,500 to 9,000 plant species are found in the region, approximately 70 percent of which are endemic. Unfortunately, nearly 1,700 of these species are threatened or endangered, and dozens have already gone extinct (Droop 1999; Low and Rebelo 1996; Conservation International 2002).

The primary threat to flora in the Cape region is invasive plant species. Pine trees from California and Europe have replaced the scrubby native vegetation in many places. Agriculture has also destroyed large areas of habitat. Human development has consumed a third of all flora, mostly in lowland areas, where 80 percent of the habitat has been lost. Urbanization also threatens endemic species near cities. In fact, the greater Cape Town area contains the highest concentration of threatened species in the world (Cowling et al. 1997).

Desert and Semidesert

True deserts cover a large portion of Africa and the Middle East. The largest desert in the world, the Sahara, rivals the continental United States in total land area. Despite its size, however, it harbors little biodiversity, and of its 1,600 plant species, only 10 to 15 percent are endemic (James 1993). Nevertheless, several large ungulates inhabit the Sahara and, to a larger extent, the semidesert Sahel to the south. Endemic species that are endangered or threatened include the addax, the slender-horned gazelle, and the scimitar-horned oryx. These Saharan ungulates have declined because of hunting and competition with pastoralists. Additionally, the southward march of the Sahara—caused by climate change, overgrazing, and agricultural mismanagement—has threatened Sahelian and savanna species with a reduction in habitat.

The second-largest desert in Africa, the Namib, is thought to be the world's oldest, having remained very dry for 55 million years. Over time, a large number of endemic species evolved in this area along coastal Namibia and Angola. Most large mammals in the Namib were hunted out by European settlers, but now the habitat is fairly well protected. To the east, the Namib merges with the semiarid succulent Karoo ecosystem, which is shared by Namibia and South Africa. Composed of dwarf, open shrubland, this is the richest desert in the world in terms of biological diversity. It has the highest diversity of succulent (moisture-retaining) plant species anywhere in the world. In addition, it holds 72 distinct species of lizard and 50 different species of scorpion, and 40 percent of its nearly 5,000 species of flora and fauna are found nowhere else in the world. Of these species, however, more than 600 have been classified as threatened (Wilson 2002; Conservation International 2002).

Nearly all of the Middle East region is characterized by arid and semiarid environments, but there are notable variations in elevation. This has encouraged higher levels of biological diversity, but the full extent of the Middle East's biological resources is difficult to gauge because of gaps in resource monitoring and research. The marine components in the region, such as the Mediterranean Sea, the Red Sea, and the Indian Ocean, also have high diversity, with high levels of endemism in places like Yemen's Socotra Islands. Species richness in the region varies considerably from country to country. Israel, for example, harbors 116 known mammal species, while the United Arab Emirates holds only 25. Countries with the highest numbers of threatened species in the region are Turkey (81 threatened species, including 28 species of mammals and birds) and Yemen (73 species, including 52 plant species) (World Conservation Union 2002).

Aquatic and Marine Habitats

The freshwater habitats of Africa have lost a larger proportion of species than any of the continent's other ecosystems. The Rift Valley lakes of east-central Africa house the world's richest diversity of tropical freshwater fish, with more than 800 species of cichlids (a tropical family of fish, many of which are prized for aquariums). Two hundred endemic species of fish are believed to have disappeared from the waters of Lake Victoria, and many others are on the brink of extinction because of competition and predation from the Nile Perch and Nile tilapia. Other threats to native fish include overfishing, eutrophication, sedimentation at the mouths of rivers, and the alteration of habitats by introduced plants such as the water hyacinth, which is bedeviling aquatic ecosystems in Malawi, South Africa, Tanzania, Uganda, Kenya, Zambia, and Zimbabwe (Chenje and Johnson 1994).

Development has seriously impacted biodiversity in Africa's river basins and their associated lakes and wetlands. Irrigation, in particular, has depleted water levels and altered downstream habitats. In Lake Chad, for example, the World Wide Fund for Nature reports that low water has eliminated many nesting sites for the endangered black-crowned crane. Dams also disrupt key habitat by holding water during the wet season and then releasing it during the rest of the year. Unseasonal flooding caused by a dam on the Kafue River in Zambia has threatened several other crane populations and destroyed seasonal grazing habitat for the vulnerable lechwe antelope. Dams also trap sediment and reduce the flow of nutrients to wetland and coastal ecosystems, on which many aquatic species depend. Water pollution has also harmed populations of species such as West Africa's goliath frog, the largest frog in the world.

Species-rich habitats along Africa's coasts have also suffered degradation. The continent's coral reef ecosystems, which run patchily down the east coast and contain at least three endemic species of coral, are under particular pressure. The Australian Institute of Marine Science estimates that 40 percent of East African coral reefs were destroyed in 2000 by storms, sedimentation, and dynamite fishing, and another 40 percent are threatened with destruction in the next thirty years. Mangrove forests, which house rich populations of small shrimp, fish, and birds, and serve as breeding sites for many coastal and open ocean fisheries, have also been lost in many places. In Ghana, for example, two-thirds of the country's mangrove forests were cut down in the past thirty years, according to the Global Mangrove Database and Information System. Seagrass beds are also disappearing from coastal East Africa on account of development, sedimentation, and destructive trawling activity. These environments support the endangered dugong; in 2002 the UN Environment Program reported that the dugong could soon disappear from Africa's coasts as a result of habitat loss.

Madagascar and Other Indian Ocean Islands

The islands of Comoros, Madagascar, Mauritius, and Seychelles, located out in the waters of the Western Indian Ocean, evolved in isolation from continental Africa. Moreover, colonization by humans was relatively recent. These factors, combined with climate and topography, created a wondrous array of flora and fauna that are strikingly different from those in any other part of Africa—or the world.

Madagascar is deservedly the most famous of these islands. Although geographers typically lump Madagascar with the rest of Africa, some consider it to be the world's "eighth continent" from a biological perspective. Once attached to Africa, this Texas-size island broke free 150 to 165 million years ago and drifted to its present spot off the continent's southeast coast. The island's isolated species evolved independently, inspiring the primatologist Allison Jolly to write that, on Madagascar, "time has broken its banks and flowed to the present down a different channel." Today, Madagascar has the highest number of endemic species in Africa, and it ranks sixth in the world in them. In addition, over 80 percent of Madagascar's species are endemic, the highest percentage of any major ecological region in the world. Many of these species are incomparably strange, like the "octopus trees" of the Spiny Desert; the highly endangered aye-aye, which resembles a cross between a monkey, a bat, and a woodpecker; and the giraffe-necked weevil, a red rain forest insect with a neck like a fire-truck rescue ladder (Tyson 2000).

Madagascar is perhaps best known for its lemurs, an order of primate endemic to the country and a few surrounding small islands. About one-third of all known lemur species are now extinct, but Madagascar still supports thirty-three species, four of which are classified as endangered and nine as vulnerable (World Conservation Union 2002). Today, Madagascar holds more threatened species than any other African country except Tanzania. Habitat

Black and white ruffed lemur in tree, Nosy Komba, Madagascar. GALLO IMAGES/CORBIS

loss has taken the greatest toll; humans destroyed 90 percent of the original natural vegetation, and protected areas are few and often poorly managed (Tyson 2000).

With a higher proportion of endangered species than Madagascar, the nearby Mascarene Islands have been dubbed "islands of the living dead." Like Madagascar, these islands contain many unique species, but their small size makes populations highly vulnerable to disturbance. Invasive species wreak their most pernicious effects on islands because many such species have few if any natural predators. For example, introduced monkeys on Mauritius have devastated the last remaining stand of *Elaeocarpus bojeri* trees because they snapped off all of their fruits before they could produce seeds. These trees constitute one of some fifty species of plants on the islands that survive with just one or a few remaining individuals, and many continue to go extinct before biologists' eyes.

Endemic and highly endangered bird species also inhabit the Mascarenes. One of them, the Mauritius kestrel, was the rarest falcon in the world, and some biologists argued that it was too far gone to rescue. After a well-publicized conservation campaign and a captive breeding program led by Durrell Wildlife Conservation Trust/Jersey Zoo, however, it miraculously recovered and now numbers several hundred in the wild (Yoon 1999; Durrell Wildlife Conservation Trust 2001).

Sources:

Bryant, Dirk, D. Nielson, and L. Tangley. 1997. *The Last Frontier Forests: Ecosystems and Economies on the Edge.* Washington, DC: World Resources Institute.

Burgess, Neil D., and G. Phillip Clarke. 2000. *Coastal Forest of Eastern Africa.* Norwich, UK: Page Bros.

Chenje, Munyaradzi, and Phyllis Johnson, eds. 1994. *State of the Environment in Southern Africa.* Harare, Zimbabwe: South African Research and Documentation Centre, IUCN-World Conservation Union, and Southern African Development Community.

CITES. 2003. "CITES Appendix II." Available at http://www.cites.org/eng/append/latest_append.shtml (accessed April 2003).

Conservation International. 2002. "Biodiversity Hotspots." Available at www.biodiversityhotspots.org (accessed December 13, 2002).

Cowling, R. M., D. M. Richardson, and S. M. Pierce. 1997. *Vegetation of Southern Africa.* Cambridge: Cambridge University Press.

Droop, Steven, Michael Moller, and Martin Jenkins. 1999. "The Ethiopian Region." In John A. Burton, ed., *The Atlas of Endangered Species.* New York: Quarto.

Durrell Wildlife Conservation Trust. 2001. "Mauritius Kestrel" (March). Available at http://www.durrellwildlife.org/upload/MainSite/Documents/pdfs/Mauritius%20kestrel.pdf (accessed February 2003).

Eckholm, E., et al. 1984. *Fuelwood: The Energy Crisis That Won't Go Away.* Washington, DC: International Institute for Environment and Development.

Hutton, Jon, and Barnabas Dickson. 2000. *Endangered Species, Threatened Convention.* London: Earthscan.

James, Valentine Udoh. 1993. *Africa's Ecology: Sustaining the Biological and Environmental Diversity of a Continent.* Jefferson, NC: McFarland.

Kaiza-Boshe, Theonestina, Byarugaba Kamara, and John Mugabe. 1998. "Biodiversity Management in Tanzania." In John Mugabe, ed., *Managing Biodiversity: National Systems of Conservation and Innovation in Africa.* Nairobi, Kenya: African Centre for Technology Studies Press.

Lock, J. M. 2001. "Africa, Ecosystems of." In *Encyclopedia of Biodiversity,* vol. 1. San Diego: Academic Press.

Lovett, J. C. 1998. "Eastern Arc Mountain Forests: Past and Present." In L. Schulman, L. Junikka, A. Mndolwa, and I. Rajabu, eds., *Trees of Amani Nature Reserve.* Helsinki: Helsinki University Press.

Low, A. B., and A. G. Rebelo, eds. 1996. *Vegetation of South Africa, Lesotho and Swaziland.* Pretoria, South Africa: Department of Environmental Affairs and Tourism.

Martin, Claude. 1991. *The Rainforests of West Africa: Ecology, Threats and Protection.* Basel, Switzerland: Berkhauser Verlag.

McClanahan, T. R., and T. P. Young. 1996. *East African Ecosystems and Their Conservation.* New York: Oxford University Press.

McKinley, James C., Jr. 1998. "It's Kenya's Farmers vs. Wildlife, and the Animals Are Losing." *New York Times,* August 2.

Ministry of Finance Planning and Economic Development. 2000. *Statistical Abstracts 2000.* Kampala, Uganda: MFPED.

Mittermeier, Russell A., Norman Myers, and Cristina Goettsch Mittermeier. 1999. *Hotspots: Earth's Biologically Richest and Most Endangered Terrestrial Ecoregions.* Washington, DC: CEMEX, Conservation International.

Mugabe, John, Njeri Marekia, and David Mukii. 1998. "Biodiversity Management in Kenya." In John Mugabe, ed., *Managing Biodiversity: National Systems of Conservation and Innovation in Africa.* Nairobi, Kenya: African Centre for Technology Studies Press.

Osei, William Yaw. 1997. "Human-Environmental Impacts: Forest Degradation and Desertification." In Samuel Aryeetey-Attoh, ed., *The Geography of Sub-Saharan Africa.* Upper Saddle River, NJ: Prentice Hall.

Perrings, Charles. 2000. *The Economics of Biodiversity Conservation in Sub-Saharan Africa.* Northampton, MA: Edward Elgar.

Pierce, Fred. 1994. "Soldiers Lay Waste to Africa's Oldest Park." *New Scientist* 44 (December 3).

Population Reference Bureau. "PRB 2002 World Population Data Sheet." Available at www.prb.org.

Quézel, P. F. Médail, R. Loisel, and M. Barbero. 1999. "Biodiversity and Conservation of Forest Species in the Mediterranean Basin." *Unasylva No. 197—Mediterranean Forests.* Rome: FAO.

Reading, Richard P., and Brian Miller. 2000. *Endangered Animals: A Reference Guide to Conflicting Issues.* Westport, CT: Greenwood Press.

Strieker, Gary. 2000. "Summit Laws Unable to Protect Most Endangered Species." *CNN.com* (May 11).

Tedla, Shibru, and Martha Gebre. 1998. "Biodiversity Management in Ethiopia." In John Mugabe, ed., *Managing Biodiversity: National Systems of Conservation and Innovation in Africa.* Nairobi, Kenya: African Centre for Technology Studies Press.

Tesi, Moses K. 2000. *The Environment and Development in Africa.* Lanham, MD: Lexington.

Tyson, Peter. 2000. *The Eighth Continent: Life, Death, and Discovery in the Lost World of Madagascar.* New York: William Morrow.

UN Environment Programme. 1999. *Global Environmental Outlook 2000.* London: Earthscan.

———. 2002. *Africa Environment Outlook.* Hertfordshire, UK: Earthprint Limited and UNEP.

UN Food and Agriculture Organization. 2001a. *Global Forest Resources Assessment 2000.* Rome: FAO.

———. 2001b. *State of the World's Forests 2001.* Rome: FAO.

UN Secretariat of the Convention on Biological Diversity. 2001. *Global Biodiversity Outlook.* Montreal: Productions MR.

Verrengia, Joseph B. 2001. "As Poachers Purge Jungles, Species Face Extinction." *Los Angeles Times,* July 23.

Wilson, E. O. 2002. *The Future of Life.* New York: Alfred A. Knopf.

World Conservation Monitoring Centre. 1992. *Global Biodiversity: Status of the Earth's Living Resources.* London: Chapman and Hall.

———. 2000. *Global Biodiversity: Earth's Living Resources in the 21st Century.* Cambridge, UK: World Conservation Press.

World Conservation Union (IUCN). 1990. *Biodiversity in Sub-Saharan Africa and Its Islands.* Gland, Switzerland: IUCN.

———. 1998. *The Congo Basin: Human and Natural Resources.* Amsterdam: Netherlands Committee for IUCN.

———. 2000. *2000 IUCN Red List of Threatened Animals.* Gland, Switzerland: IUCN.

———. 2002. *2002 IUCN Red List of Threatened Species.* Gland, Switzerland: IUCN. Available at www.redlist.org (accessed December 12, 2002).

World Resources Institute. 1992. *World Resources: 1992–1993.* New York: UNDP and World Bank.

Yoon, Carol Kaesuk. 1999. "Watching as Species Fall into Extinction." *New York Times,* March 16.

Parks,
Preserves, and
Protected Areas

Africa—including sub-Saharan Africa as well as North Africa and the Middle East—contains more than 1,800 internationally recognized protected areas that cover approximately 3.1 million square kilometers. Several of the parks and preserves in the region rank among the world's largest protected areas. In North Africa and the Middle East, for example, Ar-Rub'al-Khali Wildlife Management Area in Saudi Arabia ranks as the second-largest protected area in the world at 640,000 square kilometers. The Northern Wildlife Management Zone, also in Saudi Arabia, is the world's sixth-largest protected area at 100,875 square kilometers (Greene and Paine 1997). Southern Africa also contains its share of large protected areas, including the Selous Game Reserve in Tanzania, at 52,200 square kilometers; Namib-Naukluft National Park in Namibia, at 49,768 square kilometers; and the Okavango Delta in Botswana, which is the largest inland delta in the world, at 16,000 square kilometers (McCullum 2000).

Despite their size and abundance, however, Africa's protected areas face serious threats as the region struggles to meet the basic needs of its large and rapidly growing human population. Some of the most common problems affecting protected areas include deforestation, conversion of land for agriculture, incursions by livestock onto limited grazing areas, encroachment of human settlements, and poaching of wildlife. In addition, political instability and a lack of financial resources have led to inadequate infrastructure and enforcement in many of Africa's protected areas. Experts predict that many of the region's parks and preserves will suffer severe degradation or even disappear in the twenty-first century without international support for programs that involve local communities in resource management and create economic opportunities for local people.

Classification of Protected Areas

Protected areas around the world are managed for a wide range of purposes, including scientific research, wilderness protection, preservation of species and ecosystems, maintenance of environmental services, protection of specific natural and cultural features, tourism and recreation, education, sustainable exploitation of natural resources, and maintenance of cultural and traditional attributes. The specific design, objectives, implementation, and management of protected areas all vary in accordance with the home country's cultural, political, economic, and ecological orientations. Indeed, classification systems used by individual countries vary in accordance with objectives and levels of protection, and title designations are different from country to country as well. Therefore, comparing protected areas in different regions of the world, or in different countries within one region, can be a challenging task.

The World Conservation Union (IUCN), formerly the International Union for the Conservation of Nature and Natural Resources, has over the years established and modified a classification system for protected areas that is recognized around the world. Fitting protected areas into this system based on their objectives, regardless of their local designations, makes information comparable across national and regional boundaries, allowing an assessment of the effectiveness of different protected area categories. Data on all but the smallest of the world's parks and reserves are collected by the World Commission on Protected Areas and used to create the *United Nations List of Protected Areas,* the definitive listing of protected areas around the globe.

The World Conservation Union classifies each formally designated protected area in one of six management categories. Category I parks and reserves are protected areas managed primarily for science or wilderness protection. Strict Nature Reserves (Category Ia) includes ecological reserves, biological reserves, ecological stations and other areas that are managed purely for biodiversity protection and scientific research and in which nonscientific activities are generally not permitted. Wilderness areas (Category Ib) are protected areas managed primarily for wilderness ecosystem protection and tourism; they generally allow human visits only at a primitive level—that is, without assistance from man-made infrastructure such as roads and hotels.

Category II protected areas are national parks managed for both ecosystem protection and human recreation. This is the most common category of protected area everywhere . . . because it is both the oldest of the categories and the one that is best suited to achieve the two objectives of greatest interest to the general public—conservation and recreation.

Table 3.1 Protected Areas in Africa, 1999

NATIONALLY PROTECTED AREAS

		Terrestrial		Marine
	No.	Area (000 ha)	% Land area	No.
Central	69	31,161	33.1	10
Eastern	119	11,981	n/a	16
Northern	72	15,862	7.8	50
Southern	578	65,014	n/a	44
Western	123	28,724	68.2	25
WIOI	89	N/A	N/A	3
Total	1,050	N/A	N/A	148

INTERNATIONALLY PROTECTED AREAS

	Biosphere Reserves*		World Heritage Sites		Ramsar Sites	
	No.	Area (000 ha)	No.	Area (000 ha)	No.	Area (000 ha)
Central	11	3,034	7	9,121	8	4,228
Eastern	7	1,126	5	454	5	105
Northern	13	N/A	2	>13	22	>2000
Southern	8	N/A	10	7,850	27	12,026
Western	15	31,112	10	12,003	37	3,674
WIOI	3	N/A	3	N/A	4	53
Total	57	N/A	37	>29,441	103	>22,086

SOURCE: UN Environment Programme. 2002. *Africa Environment Outlook: Past, Present and Future Perspectives.* http://www.unep.org/aeo/064.htm. Accessed July 22, 2003.

*Some Biosphere Reserves are also World Heritage Sites or Ramsar sites

Other management classifications are available for natural monuments and landmarks that are managed primarily for conservation of specific natural features such as waterfalls, cave systems, or canyons (Category III), species and habitat protection areas that are managed primarily for conservation, though subject to tree felling and other active forms of management (Category IV), protected landscapes and seascapes with dual conservation and recreation management mandates (Category V), and "managed resource protection areas" (Category VI), which seek to balance biodiversity protection with extractive activities such as logging conducted in a sustainable manner.

Greater Limpopo Transfrontier Park

In late 2002, the heads of three southern African states signed a historic treaty that will eventually combine Kruger National Park in South Africa, Limpopo National Park in Mozambique, and Gonarezhou National Park in Zimbabwe into one 35,000-square-kilometer conservation area (three times the size of Yellowstone National Park in the United States) to be called Greater Limpopo Transfrontier Park. The creation of Greater Limpopo is indicative of an emerging trend toward international cooperation in the management of protected areas in Africa.

The world's first international protected area, Waterton-Glacier International Peace Park, was created in 1932 by administrators of Waterton National Park in Canada and Glacier National Park in the United States. Since then, numerous protected areas have been established on both sides of political borders, creating protected lands capable of preserving intact ecosystems of considerable size. Most of these efforts were intended to address the fragmentation of habitat and associated loss of biodiversity that often occurred when two adjacent natural areas were artificially separated from each other by political boundaries. Many had the added benefits of encouraging tourism and promoting international cooperation and peace.

The idea of combining adjacent parks in different countries came to Africa during the 1990s. In 1997 retired South African industrialist and award-winning environmentalist Anton Rupert established the Peace Parks Foundation to raise funds and promote regional conservation projects. The foundation developed a list of eight potential sites in ten African countries and began working with the national governments to combine them.

The group's first successful effort was the creation of Kgalagadi Transfrontier Park in 1998. This park, which combined South Africa's Kalahari Gemsbok National Park and Botswana's Gemsbok National Park into a single ecological unit, allows for free migration of species across the Kalahari Desert. It also saw a threefold increase in tourism, to 150,000 visitors, in its first year of existence (Grunwald 2002).

The creation of Greater Limpopo Transfrontier Park marked another success in the "peace parks" initiative. The newly created park is expected to help reduce the fragmentation of habitat, restore wildlife to areas where it once roamed, and preserve biodiversity. The participating nations also hope that it will boost tourism, reduce costs by allowing unified park management and enforcement systems, and promote harmony between their countries.

But critics claim that Greater Limpopo is a grand idea that stands little chance of being fully implemented. They note that the parks are separated by miles of fence

(continues)

that must be torn down. They also point out that the Mozambique side has no tourist facilities and is home to 20,000 impoverished villagers who may need to be resettled. Finally, they question the participation of Zimbabwe, which suffers from political instability following Robert Mugabe's disputed election and controversial land-reform programs. Proponents admit that the process of combining the parks may take years, but they argue that the benefits will justify the work involved.

Sources:

2001."Game without Frontiers: Southern Africa's National Parks Are Linking up over National Borders to Create Vast Peace Parks." *Time International* (May 14).

Grunwald, Michael. 2002."A Grand Vision Fenced in by Reality." *Washington Post*, December 9.

Marshall, Leon. 2000."Southern Africa's Peace Parks Are Causing Excitement for Its Conservation, Tourism, and Economic Implications." *Knight Ridder/Tribune News Service*, July 5.

Protected Areas in Sub-Saharan Africa

Sub-Saharan Africa is a vast area that features a number of diverse ecosystems, from grasslands and deserts to biologically rich rain forests. Although the traditions of many of the region's native inhabitants included systems for the protection of nature, the first modern protected areas were established by European authorities under colonial rule. Kruger National Park in South Africa was established by British colonial powers in 1898, for example, while Virunga National Park in Zaire was created by Belgian colonial rulers in 1925. One of the first notable pieces of legislation concerning protected areas was the London Convention of 1933. It was superseded following independence by the African Convention on the Conservation of Nature and Natural Resources of 1968, an agreement creating a framework for defining conservation areas that was adopted by three-quarters of the states in sub-Saharan Africa.

Many African states established their own protected areas in the years following their independence from colonial rule. These early protected areas often centered around specific targets and tended to be located in remote, sparsely populated, and relatively pristine areas. As a result, the setting aside of land initially did not cause much conflict with local people. When these early protected areas were later expanded, however, their boundaries often enclosed nearby villages, displacing local people or preventing them from practicing their traditional land uses. For example, Niokolo Koba National Park in Senegal was expanded seven times over the years. Some members of the displaced communities refused to abide by the area's protected status and engaged in extensive poaching of the park's wildlife.

Protected Areas Coverage

The fifty countries of sub-Saharan Africa contained 1,270 protected areas (including more than 1,000 areas protected at the national level) in IUCN categories I through VI at the close of the 1990s. These parks and preserves cover 2.074 million square kilometers, or about 9 percent of the region's total land area. The countries of eastern and southern Africa tend to have the highest percentage of land under protection, with an average of 12.24 percent, though many of these areas fall in the higher IUCN categories and thus serve multiple-use functions. Zambia leads the way with 30.09 percent of its land protected, followed by Tanzania with 27.95 percent and Uganda with 20.78 percent. In addition, Botswana, Ethiopia, Malawi, Namibia, and Zimbabwe all protect between 10 and 20 percent of their total land area. Coverage is not uniform throughout the region, however; Lesotho and Somalia each protect less than 1 percent of their total land area (Greene and Paine 1997; UN Environment Programme 2002).

The countries of western and central Africa set aside an average of 5.66 percent of their total land as protected areas. Once again, coverage is not uniform: protected areas cover more than 10 percent of the land in Benin (11.21 percent), Burkina Faso (10.42 percent), Rwanda (15.06 percent), and Senegal (11.4 percent), but less than 2 percent in Djibouti, Guinea, Liberia, and Mauritania (ibid.).

In addition to nationally designated protected areas, many countries in sub-Saharan Africa contain areas that are protected under international conventions. More than half of the countries in the region participate in the World Heritage Convention, a UN program intended to define and conserve sites of great natural or cultural importance. As of 2000, thirty-seven World Heritage sites had been designated in the region, covering 30 million hectares (UN Environment Programme 2002). Tanzania had six World Heritage sites, the Democratic Republic of Congo contained five, and South Africa and Senegal held four sites each.

Slightly fewer countries participate in UNESCO's Man and Biosphere Program, which is intended to promote conservation and sustainable use of natural resources. As of 2002, sixty Biosphere Reserves were established in twenty-eight countries in Africa, with forty-six of these reserves located in sub-Saharan Africa. Kenya led the way with five reserves, while Guinea and South Africa each had four (UN Educational, Scientific and Cultural Organization 2002). Finally, as of 2003, thirty-three countries in sub-Saharan Africa participated in the Convention on Wetlands of International Importance, more commonly known as the RAMSAR Convention. A total of 111 sites have been designated in the region (up from 103 in 1999), with the most sites in South Africa (17) and Algeria (13) (RAMSAR 2003).

The protected areas of sub-Saharan Africa cover a wide range of habitat types, yet some gaps in coverage exist. Some of the habitats that are poorly represented in the current systems of national and international protection include freshwater lakes and swamps, grassland ecosystems, savanna and desert ecosystems, mountain ecosystems, lowland forests, coastal forests, coastal wetlands, mangroves, sand dunes, and coral reefs.

Countries of the region differ greatly in their emphasis on protecting representatives of various habitat types. Some countries have adopted a "National Systems Plan" approach and conducted inventories of their major ecosystems and species to create protected area systems. For example, Malawi's system of protected areas and forest reserves includes representatives of its major biotic communities, while Namibia's state-owned lands contain eleven of the country's fourteen major vegetation zones. Other nations that have performed well in terms of protecting a variety of habitat types include Cameroon, Côte d'Ivoire, Democratic Republic of Congo, Kenya, Senegal, Zambia, and Zimbabwe. But many other countries have neither surveyed their resources nor created plans for protecting them. In such cases, any protected areas that exist were generally established to preserve specific sites of aesthetic value.

Trends in the Creation of Protected Areas

Both the number of protected areas and the total land area covered increased sharply in sub-Saharan Africa during the 1960s and 1970s. This growth slowed but remained steady during the 1980s, then slowed considerably in the 1990s. Still, a number of significant protected areas have been established in recent years, most notably the 2002 announcement that Gabon, which previously had no national parks at all, was setting aside 10 percent of its total land area in thirteen new national parks, including several forest areas of exceptional faunal and floral biodiversity. Other protected areas of recent vintage include the Bayanga Forest in the Central African Republic, which protects 32,000 hectares. This area was created as a "third-generation" protected area, wherein conservation is considered alongside the needs of local people for economic development. Nigeria's Afi Mountain Wildlife Sanctuary, on the other hand, was established in 2000 primarily to help protect the Cross River Gorilla, drills, chimpanzees, and other primate species.

Another important though underutilized component of sub-Saharan Africa's protected area networks is marine protected areas such as Banc d'Arguin National Park on the coast of Mauritania. This important West African fishing ground enjoys national protection as well as international

Red hartebeests grazing in Salt Pan, Kalahari Gemsbok Park, Botswana. PETER JOHNSON/CORBIS

recognition as a World Heritage Site. The park consists of a huge area of shallows where cold, nutrient-rich waters rise to the ocean surface and are warmed by the sun, creating feeding and breeding grounds for numerous species of fish and migratory birds. In fact, 2 million European wading birds frequent the area each winter. The park itself is closed to fishing, but industrial trawlers ply the waters just outside its boundaries. This situation has created conflict with local people who try to make a living by fishing from small boats (Pearce and Murphy 2001).

An emerging trend in the late 1990s involved regional cooperation in the formation and management of protected areas. One direct result of this new approach is the formation of several large, transborder parks by formally merging protected areas in two or more different countries. The first southern African park to span international borders was Kgalagadi Transfrontier Park, a 14,675-square-mile park that was established in 1998 through the union of South Africa's Kalahari Gemsbok National Park and Botswana's Gemsbok National Park. The park is run by a joint management committee from the two nations. Some other examples of cross-border protected areas include Maloti-Drakensberg Park, between Lesotho and South Africa, and Greater Limpopo Transfrontier Park, between Mozambique, South Africa, and Zimbabwe (see sidebar on page 58).

Central Africa's network of protected areas has been expanded in recent years through the creation of new parks and the linking of existing ones. For example, Sangha Park was created to provide a link between Lobeke National Park in Cameroon, Dzanga-Sangha in the Central African Republic, Nouabale-Ndoki Park in Congo, and several production forests and hunting zones that surround these protected areas.

Threats to Protected Areas

Most of the threats to protected areas in sub-Saharan Africa stem from larger political, economic, and social problems affecting the region, such as rapid population growth, widespread poverty, food shortages, and political instability. These and other problems have often overwhelmed the resources of many national governments, leaving few institutional resources for the creation and management of protected areas. At the same time, the need to sustain a large and growing human population created a trend toward finding alternative uses of land. For example, large areas of wetlands were reclaimed for food production at the expense of migratory birds and other species that depended on such natural areas. In addition, huge numbers of wild game were killed to supply bushmeat to city residents, and vast tracts of forest were destroyed to provide fuelwood and charcoal.

POPULATION GROWTH

As Africa's human population continues to grow, it will only place further stress on the continent's natural resources and ecosystems, including those in protected areas. "National parks, forest reserves, and other protected areas permanently block off vast areas of land, therein forgoing competing economic uses, and often excluding human settlements," one expert explained. "Considering that a large portion of the African continent is either desert, arid or semiarid in nature, drought is frequent and recurring, and that the human population increase in Africa has been one of the highest in the world, it becomes immediately apparent why the conservation of protected areas is a serious front-line socio-economic issue" (McNeely 1994).

More than half of Africa's original wildlife habitat has already been lost as a result of deforestation, conversion of land to agriculture, and other human impacts. Although this makes parks and preserves even more important as core areas for supporting biodiversity and ecological processes, it also increases the external pressures to develop the economic potential of protected areas. Experts are currently examining the role of protected areas in sub-Saharan Africa to find ways to combine conservation and local development. Many feel that ensuring the survival of existing protected areas and encouraging the formation of new ones will mean moving toward landscape-level planning and community-based resource management. As a result, more protected areas may be required to support multiple uses in the future.

INEFFECTIVE MANAGEMENT AND ENFORCEMENT

Many protected areas in sub-Saharan Africa appear to exist mostly on paper. Although they have been formally established through legislation, they lack management plans, infrastructure, and even scientific data on which to base decisions. In some cases, management of protected areas is inadequate because responsibility is divided between a number of government agencies, creating conflicts of interest. In other cases, political instability has eviscerated a country's capacity to effectively manage its protected areas and other natural resources. For example, Sudan has officially designated 10 million hectares, or 4 percent of its land area, for parks and preserves. Yet the ongoing civil war in that country has stymied any attempt to impose meaningful protections or management schemes on these protected areas, so they remain an abstraction rather than a reality. Finally, the management of some protected areas in the region is ineffective because park managers lack technical expertise and training opportunities.

A related threat to protected areas in sub-Saharan Africa is poor enforcement of boundaries and regulations. In many cases, enforcement suffers from a shortage of manpower because of a lack of funds to pay for salaries and training. In Congo, for example, the staffing ratios in the national parks are as low as one guard per 35,000 hectares. Some countries are also seeing significant depletion of trained staff as a result of the AIDS pandemic. The protected areas of Central Africa have suffered significant degradation—in the form of encroachment for agriculture, bushmeat poaching, logging, and oil exploration—as a result of poor enforcement. In some cases, such encroachment takes place with the approval of the national government. For example, in past years Gabon has granted forest concessions to timber companies in some of its most prized reserves. Even a World Heritage Site like Taï National Park in Côte d'Ivoire, with its large tracts of undisturbed rain forest, is threatened by slash-and-burn agriculture, poaching, and illegal logging and mining activities. The international environmental organization World Wide Fund for Nature (WWF) is currently working with local communities to develop and implement a long-term management plan for the park (UN Environment Programme 2002).

INADEQUATE FINANCING

Few of the protected areas in sub-Saharan Africa are self-financing; instead, they usually represent yet another drain on already limited national resources. The majority of countries in the region have an annual protected areas budget that is less than one-fifth of the amount generally considered necessary for effective conservation. Of course, the level of funds dedicated to protected areas administration varies widely between countries. The parks of Kenya, Namibia, Rwanda, and South Africa tend to be fairly well financed, for example, while those of Angola and Sierra Leone are starved for funds. A lack of funds contributes directly to the problems of inadequate staff levels, management, and enforcement.

At the present time, relatively little protected land in the region is owned by private parties or nongovernmental organizations (NGOs); South Africa is a notable exception. A number of countries depend on investment from international agencies to keep their parks up and running, though the level of support varies from country to country. Some experts contend that long-term international financing will be necessary to support planning, development of

---

Gabon's Newly Minted National Park System

The forests, savannas, and marshes of Gabon are among Africa's major remaining bastions of wildlife habitat and biodiversity, but as the twentieth century drew to a close, the prospects for their long-term survival looked bleak. Mining and farming activities were expanding into previously undeveloped areas, and rapid acceleration of logging activity threatened large tracts of Gabonese rain forest. Even the country's handful of protected areas were not immune from these incursions. For example, logging concessions were granted in such species-rich areas as the La Lopé Wildlife Reserve, the Wonga Wongé Presidential Reserve, and the Mokalaba Reserve (Global Forest Watch 2000).

Even as these troubling developments came to pass, however, international conservation NGOs worked feverishly to reverse the nation's slide toward unsustainable plundering of its natural resources. Groups including the World Conservation Society, Conservation International, and the World Wide Fund for Nature spent years documenting the country's wildlife resources, analyzing the ecological impact of logging, hunting, farming, and other human activities, and lobbying the Gabonese authorities to consider more sustainable approaches to resource exploitation.

In 2000 these conservation efforts received a major boost when researcher Michael Fay completed a fifteen-month expedition of 2,000 miles (3,200 kilometers), made on foot through the heart of the region. This scientific foray, sponsored by the World Conservation Society and National Geographic, confirmed the rich biological wealth contained in the Gabonese section of the Congo basin. In addition, photographs and video footage gathered during the journey helped convince Gabon's president El Hadj Omar Bongo of the need for a national

(continues)

park system. "The president had heard all about national parks, sustainable development, reserves, and so on, but until he saw the imagery with his very own eyes, he didn't realize his country had resources beyond timber, manganese, and oil," recalled Fay (Mayell 2002).

In September 2002 the Gabonese government declared its intention to formally protect 10 percent of its total land area, including some of the most pristine tropical rain forests remaining on the planet. Under the plan, a total of 10,000 square miles (26,000 square kilometers) will be set aside in thirteen separate national parks that represent a cross-section of ecosystem types, from coastal areas to deep jungle. The national park plan even grants a reprieve to places like La Lopé Wildlife Reserve, which had been scheduled for logging, by incorporating them into various parks. All told, the parks will preserve essential habitat for buffalo, rhinoceros, gorillas, chimpanzees, forest elephants, and other wildlife, as well as numerous endemic plant and bird species.

Environmental groups hailed the announcement as a signal victory for Central African wildlife and a bold step that sets a new standard for African conservation. These same organizations are now moving forward to help Gabon erect the necessary infrastructure—such as trained staff and facilities to accommodate tourists—to maintain and protect the new parks, and to assist the Gabonese government in proving its assertion that ecotourism can be a viable economic alternative to logging and mining.

Conservationists acknowledge that Bongo's decision to support a national park system is a courageous one, for Gabon will lose timber revenues from logging concessions that are going to be canceled or scaled back. But while the Gabonese president admitted that the dramatic expansion of the nation's protected areas network "implies certain sacrifices in the short- and medium-term," he strongly defended the goal of "preserving these natural wonders for future generations. . . . By creating these national parks, we will develop a viable alternative to simple exploitation of natural resources that will promote the preservation of our environment. Already there is a broad consensus that Gabon has the potential to become a natural Mecca, attracting pilgrims from the four points of the compass in search of the last remaining natural wonders on earth."

Sources:

Gabon National Parks. 2002. "African Nation of Gabon Establishes National Park System." Gabon National Parks website, September 4. Available at http://gabonnationalparks.com/gnp-home (accessed February 2003).

Global Forest Watch. 2000. *A First Look at Logging in Gabon.* Washington, DC: Global Forest Watch.

Mayell, Hillary. 2002. "Gabon to Create Huge Park System for Wildlife." *National Geographic Online,* September 4. Available at http://news.nationalgeographic.com (accessed February 2003).

Quammen, David. 2001. "In the Green Abyss." *National Geographic* 199 (March).

infrastructure, hiring and training of personnel, and projects to decrease population pressure in the region's protected areas.

CONFLICT WITH TRADITIONAL LAND USES

Many parks and preserves in sub-Saharan Africa were created in areas where local communities enjoyed a long tradition of using the land for subsistence, whether through farming, grazing, hunting, or gathering of water or fuelwood. When land is placed off-limits to such use—often without consulting with traditional users or providing them with alternative means of subsistence—it is little wonder that some people grow resentful. "The extensive African experience of local communities living in harmony with their environment has generally been acknowledged and well-documented," one expert noted. "Until recently, however, there was little evidence that competent authorities took local communities and their experiences into their confidence by giving them a participatory role in the management of national parks and protected areas. As a rule, no stake in the direct economic returns accruing from the use of such protected areas was offered or forthcoming to these communities. As a result, antagonistic relationships have often developed between park authorities and local communities" (McNeely 1994).

In some instances—particularly during times of drought or political instability—local people ignore protected area boundaries and regulations and simply move in and begin using the land. For example, half of the protected areas in Tanzania have been impacted by local people settling on the land or using it for agriculture. In addition, a recent study found 125,000 square kilometers of farmland within the boundaries of protected areas in the Great Lakes Region of Central and Eastern Africa (Muster 2000). In fact, problems such as poaching, forest destruction, and encroachment have been reported in almost every country in the region. Faced with imperatives for short-term survival, local people are often unable to recognize the long-term benefits offered by protected areas. Other members of local communities are aware of these benefits, but still feel forced to engage in environmentally destructive behavior to provide for their families.

The conflict between protected areas and traditional land users is expected to become more pronounced as the human population of the region continues to grow. This conflict may be the most important challenge to conservation in sub-Saharan Africa. Experts emphasize that turning over protected lands to local communities would provide only a temporary solution, because the lands would soon become degraded by overuse. Instead, they recommend that protected areas be managed cooperatively with affected communities, incorporating multiple-use areas and compatible development in buffer zones surrounding parks.

POLITICAL INSTABILITY

Political instability and war harm protected areas in a number of ways. For instance, military activities can cause direct damage to the land, wildlife, and infrastructure in parks and preserves. Four national parks in Ethiopia lost all their facilities, including ranger camps and equipment, during that country's civil war. In Mozambique, the civil war caused significant loss of habitat and species in Gorongosa National Park and Marromeu Buffalo Reserve in the Zambezi Delta (Chenje 2000). Another way in which political instability and war impact protected areas is through the creation of refugees. In many cases, displaced people have formed settlements inside parks, contributing to the degradation of vegetation, water, and wildlife. Political instability also frequently leads to a collapse in management and enforcement of protected areas.

HABITAT FRAGMENTATION

Isolation is another threat facing the protected areas of sub-Saharan Africa. For example, the total forest area in Nigeria has decreased dramatically in the past quarter-century, creating a network of isolated forest reserves surrounded by human settlements and various forms of development. Incompatible human land uses increasingly surround parks and preserves, cutting off migration routes and creating "islands" in which species become vulnerable to local extinction.

Several nations have attempted to address the isolation of habitats by working cooperatively with their neighbors to create cross-border parks. These large parks are formed by combining adjacent protected areas in two or more countries to establish migration corridors for wildlife. Other benefits often include reduced administration costs and increased tourism. Greater Limpopo National Park—which was established in 2002 through the merger of parks in Mozambique, South Africa, and Zimbabwe, as well as several private game and forest reserves in the area—now ranks among the largest protected areas in the world (see sidebar on page 58).

Local concerns also affect a number of protected areas in sub-Saharan Africa. Commercial fishing threatens many marine preserves, for instance, while development activities, invasion of exotic species, and tourist pressure threaten many areas. On the plus side, the region's protected areas (with the exception of a few lakes and coastal and marine protected areas) see little damage from pollution, compared with parks in more industrialized areas.

Protected Areas in North Africa and the Middle East

The region encompassing North Africa and the Middle East has historically been one of the planet's major crossroads for immigration and commerce. It is located at the junction of three continents—Africa, Asia, and Europe—and borders many of the world's great seas, including the Arabian Sea, Black Sea, Caspian Sea, Mediterranean Sea, and Red Sea, as well as the Atlantic and Indian oceans. The region exhibits great variation in elevation and precipitation, and its terrain ranges from desert plains to broad-leafed forests. As a result of its geographic location and wide range of habitats, North Africa and the Middle East have nourished their own unique biodiversity. At the same time, however, most of the region's natural areas have been impacted by human activity. A great deal of degradation has occurred during the past fifty years, caused by human population growth and the resulting demand for increased economic development.

The countries of North Africa and the Middle East show enormous differences in the amount of land that they protect as well as in the effectiveness of their protected areas legislation, management, and enforcement. Fewer than one-third of the countries in the region have well-implemented land-protection programs, and some countries provide no national protection at all. As in sub-Saharan Africa, many countries in the region are politically unstable and suffer from a variety of social and economic problems. As a result, the establishment of protected areas is often given low priority, and few countries have undertaken scientific surveys of their natural heritage (Llewellyn 2000).

Nevertheless, the region has a long history of conservation. There is some evidence that the Greeks and Romans established the first protected areas in the region more than 2,000 years ago. Some of the ancient traditions of land stewardship and preservation—which developed as people first recognized the need to conserve scarce resources—persist today. For example, the Bedouin tribal tradition of cooperative management of grazing areas— known as *hema*—is still practiced on the Arabian Peninsula. The *hema* system is one of the world's oldest methods of ensuring the protection and efficient use of land. Under *hema,* certain areas are set aside for use during periods of drought; grazing, hunting, and other human uses are restricted in these areas (ibid.). A 1969 survey found 3,000 *hema* cooperatives in North Africa and the Middle East, though another survey conducted fifteen years later found only seventy-one in the mountain areas west of Saudi Arabia (Sulayem 2002).

The first modern protected areas in North Africa and the Middle East were established in the early twentieth century. Many of these early areas were intended more for recreation than for nature conservation. The number of protected areas in the region grew steadily from the 1950s through the 1980s. For example, between 1965 and 1977 Iran expanded its protected areas from eleven sites covering 600,000 hectares to sixty-nine sites covering nearly 8 million hectares (McNeely 1994). By the early 1990s, nearly all countries had established some protected areas. The total area under protection expanded rapidly during the 1990s because of the creation of a few large parks and preserves.

Although economic development remained the top priority for many nations, environmental awareness began to spread through parts of the region. In fact, several countries have signaled intentions to further increase the extent of their protected areas in the near future, including Lebanon, Syria, and Turkey; Egypt has expanded its protected area network considerably in the last few years.

Protected Areas Coverage

The nineteen countries in North Africa and the Middle East contain 542 protected areas in IUCN categories I through VI, covering an estimated 1.038 million square kilometers, or 8.06 percent of the region's total land area. Around two-thirds of these protected areas fall into IUCN Category VI, meaning that they are managed for sustainable use of resources. Only three countries protect more than 10 percent of their total land area: Saudi Arabia (34.39 percent); Israel (15.68 percent); and Oman (12.61 percent). A number of countries protected less than 1 percent of their total land area in the late 1990s, including Egypt, Iraq, Lebanon, Morocco, Qatar, Tunisia, and Yemen (Greene and Paine 1997). It should be noted, however, that countries such as Egypt have increased the size and number of parks and reserves since that time.

It is somewhat difficult to obtain an accurate count of protected areas in North Africa and the Middle East because many of the designations used to categorize protected areas in the region differ from those used in other parts of the world. The most common national designations in the mid-1990s included nature reserve (which accounted for 19 percent of designated protected areas), recreational area (18 percent), protected area (9 percent), game reserve (6 percent), national park (4 percent), and breeding station (3 percent). Most of these sites are small and support multiple uses; only 19 percent are considered protected areas under IUCN's classification scheme. Preserving biodiversity is the primary function of 47 percent of the region's protected areas, followed by recreation (20 percent) and hunting or game areas (13 percent) (McNeely 1994).

In addition to nationally designated protected areas, many countries in North Africa and the Middle East contain areas that are protected under international conventions. Nearly all the countries in the region participate in the World Heritage Convention. In addition, twelve countries have signed the RAMSAR Convention, under which they have protected more than sixty wetlands of international importance, including twenty-one in Iran alone. Seven countries participate in the Biosphere Reserves program, with a total of twenty-three sites in the region. Some countries also take part in regional protection efforts, such as the Mediterranean Specially Protected Areas or Council of Europe Biogenetic Reserves.

Aside from nationally and internationally designated protected areas, many countries in North Africa and the Middle East contain natural areas that receive de facto protection. These areas may enjoy such protection through religious tradition (for example, sacred groves or sites such as Mt. Sinai in Egypt) or tribal designation, or because they contain underground reserves of oil or water. In the latter case, the national government often owns such land and restricts grazing, development, and other activities, which helps the land to become valuable habitat for wildlife.

The protected areas of North Africa and the Middle East cover a wide range of habitat types, yet some gaps in coverage exist. Some of the habitats that are poorly represented in the current systems of national and international protection include the wetlands of Iraq and Afghanistan and the mountain coniferous forests of Cyprus, Lebanon, and Syria. There is also a strong need for greater protection of coastal and marine environments, which foster rich biodiversity but face threats from pollution and tourism.

The countries of North Africa and the Middle East vary in the emphasis they place on protecting various habitat types. A number of countries in the region have not yet developed comprehensive plans for the establishment of protected areas. Furthermore, only half of the countries in the region have conducted national surveys of their flora and fauna, making it difficult to create conservation plans for the entire region. Finally, no regional conservation network exists for North Africa and the Middle East, which also works against efforts to ensure that protected areas cover all of the region's various habitat types.

Threats to Protected Areas

As in sub-Saharan Africa, many of the threats facing protected areas in North Africa and the Middle East stem from the larger political, economic, and social problems affecting the region. Confronted with increasing pressure for economic development to support a growing human population, along with a lack of financial resources to dedicate to parks, many countries have placed a low priority on establishing and maintaining protected areas. This situation has led to ineffective management and law enforcement in protected area networks, which in turn has led to degradation through inappropriate human use and activities.

INADEQUATE FUNDING AND INEFFECTIVE MANAGEMENT

Most protected areas in North Africa and the Middle East are controlled by national governments. In many cases, however, the responsibility for protected area management is divided between several government agencies. This situation often leads to poor coordination of efforts and conflicts of interest.

Compounding this problem in many countries is a shortage of funds for protected areas management and enforcement. Most protected areas require government funds for their operation. Only a few parks and preserves generate enough money to cover their own costs. For example, Ras Mohammed National Park in Egypt has attracted enough tourism revenue to create jobs and a thriving economy for nearby villages.

The budgets for protected areas vary tremendously between nations. Most countries in the region receive limited financial assistance from international agencies or private investors. Some have experienced a degree of success in generating income for protected areas management by charging park entry fees or by using innovative programs such as placing surcharges on airline tickets. Partly because of the differences in budgets between countries, the number of staff assigned to protected areas varies greatly throughout the region. In general, cultural areas tend to receive greater allotments of money and personnel than natural areas.

POLITICAL INSTABILITY

Political instability and armed conflict have made both positive and negative impacts on protected areas in North Africa and the Middle East. In a few cases, long-standing enmity between states has encouraged the formation of large, undisturbed natural areas along national borders. For example, the Gebel Elba area forms a natural barrier between Egypt and Sudan. In other cases, however, armed conflict has caused terrible devastation of natural areas. For example, wars in Afghanistan have damaged the environment through the abandonment of land, uncontrolled hunting of wildlife, destruction of forests, and planting of land-mines. The 1991 Persian Gulf War caused both direct and indirect damage to protected areas in Iraq and Kuwait, including turtle nesting beaches and bird sanctuaries.

ISOLATION OF HABITATS

Many protected areas in North Africa and the Middle East are small and surrounded by areas of heavy human use. Such fragmentation of habitat often leads to a loss of biodiversity. Experts recommend expanding protected areas and establishing migration corridors between different areas in order to ensure the survival of large desert ungulates.

HUMAN IMPACTS

The protected areas of North Africa and the Middle East have suffered degradation as a result of human impacts, particularly during the past fifty years. For example, a number of wetland preserves are threatened by a lowering of the water table caused by the demands of a growing human population. In addition, large areas of steppe and mountain pasture ecosystems are being converted for agriculture and livestock grazing. Forests are also being destroyed for timber and fuelwood; this problem poses a particular threat to parks and preserves in Algeria, Cyprus, Lebanon, Morocco, and Syria. Furthermore, some protected areas are losing wildlife to unregulated hunting. Some sources estimate that between 15 and 20 million birds—including many protected species—are killed each year by the region's hunters.

As North Africa and the Middle East have become more industrialized, more protected areas have been impacted by air pollution. For example, some parks and preserves in western Turkey are suffering from the effects of acid rain. In addition, water pollution threatens some coastal and marine protected areas in the region—particularly those located near industrial centers—and some have suffered degradation as a result of oil spills. Finally, development and tourism pose threats to several protected areas, especially in the Mediterranean region. In fact, the construction of roads, hotels, and other tourist facilities has been called the most serious threat to natural areas in Cyprus, Egypt, Morocco, Tunisia, and Turkey.

Addressing Threats to Protected Areas in Africa and the Middle East

Although the protected areas of Africa and the Middle East face an array of serious and diverse threats—and many currently suffer from gradual degradation—some experts view it as a positive sign that no large areas in the region have been formally removed from protected status. To some extent, the long-term viability of the region's parks and preserves will depend upon international support for Africa's economic recovery. This process will involve such difficult tasks as diversifying economies, reducing dependence on a small range of raw materials, and restructuring the pricing of commodities to reflect their real cost.

Many experts believe that the other major factor in ensuring the future survival of protected areas will be addressing the needs of local communities and involving them in the management of parks and preserves. In fact, some experts claim that the political, economic, and social problems entrenched in Africa and parts of the Middle East make the conventional conservation model obsolete. They say that the region's growing population and demand for economic development means that the creation of large, pristine, remote parks will no longer be feasible in many countries (CARPE 2001).

An emerging school of thought on protected areas management in Africa in particular says that the majority of the continent's parks and preserves should be multiple-use areas that allow limited exploitation of natural re-sources on the basis of scientific criteria. It also favors developing regional conservation strategies that incorporate the traditional knowledge and prac-tices of local people and provide local communities with alternative resources. According to this theory, protected areas should be managed in a way that would give greater responsibility and benefits to local people. For example, rather than instituting strict hunting bans, park regulations could distinguish between traditional and modern hunting, establish open and closed seasons, provide incentives for private landowners to allow wildlife on their property, and encourage local people to breed game (McNeely 1994).

A number of programs have shown some success in encouraging the sus-tainable use of natural resources in and around protected areas. The Community Based Natural Resource Management Programs (CBNRM) that have been implemented in Kenya, Uganda, and Zimbabwe offer one example (Hulme and Murphree 2001). In Ghana, recent conservation efforts have cen-tered around reintroducing the traditional grove system of land management in order to preserve biodiversity and encourage sustainable use of resources (Oteng-Yeboah 1996). Syria has also recently returned some protected land to the control of traditional tribal cooperatives.

Some experts believe that private game reserves might be another tool to be used in support of protected areas. For example, South Africa employs a so-phisticated system of private land management to create buffer zones around protected areas. Several countries in the Middle East rely on the assistance of voluntary conservation groups to help them implement conservation policies in a cost-effective manner; examples include the Society for Protection of Nature in Israel and the Royal Society for Conservation of Nature in Jordan.

Some other successful conservation programs have focused on giving local people a direct role in protected area management. For example, South Africa's Working for Water project combines conservation with poverty alle-viation by hiring the rural poor to remove alien plant species from parks and preserves. This program provides temporary employment to around 20,000 people in more than 300 projects across the country, including about 600 peo-ple who work to remove invasive prickly pear from Kruger National Park. In addition to wages, local people also see other benefits from their employment: invasive tree species are cut and sold as firewood or charcoal, with proceeds going to local communities.

Ecotourism in Africa

Another conservation approach that has gained popularity in Africa over the last few decades is ecotourism, a fast-growing genre within the blockbuster

Rwandan Biologist Works to Save Country's Devastated Parks

Mountain Gorillas, Rwanda. GALLO IMAGES/CORBIS

Rwandan biologist Eugene Rutagarama has braved ethnic violence and civil war in his efforts to protect the endangered mountain gorilla and its habitat. Rutagarama was born in 1955 in southwestern Rwanda, near the border of the former Zaire (now the Democratic Republic of Congo, or DRC). As a member of the minority Tutsi ethnic group, he has faced persecution from extremists of the rival Hutu ethnic group throughout his life. Rutagarama and his family, which made a living buying and selling cattle, were first forced to flee Rwanda to escape ethnic violence in 1960. "I remember my family running from our home," he noted. "I remember which houses were burning as we left our village, and which way we took" (Snell 2001). Although they soon returned to Rwanda, they were forced to leave again in 1963, and a third time

in 1973. As a result, Rutagarama spent much of his youth in Zaire and Burundi. "I was a refugee, but I always had the hope of returning to my country—and that my country would return to itself," he stated (ibid.).

Rutagarama—who speaks Kinyarwanda, French, and English—trained as a wildlife biologist. He eventually chose to focus his research efforts on the endangered mountain gorillas of Rwanda. Only about 650 mountain gorillas survive in the wild. About half of these animals live in the Virunga mountains that straddle the borders of Rwanda, Uganda, and the DRC. This gorilla habitat is protected in three separate national parks in the three countries: Volcano National Park in Rwanda, Mgahinga Gorilla National Park in Uganda, and Virunga National Park in the DRC. "You can never say

(continues)

exactly what influenced you. But I remember studying the South African springbok, which had become extinguished, and thinking that maybe I could make a difference in my country one day and keep the gorilla from meeting the same end," Rutagarama recalled. "When I started to work with gorillas I immediately sensed that it was a privilege. But I also understood that it was a duty and a challenge. If I chose this for my work, I must succeed. Gorillas have a right to survive" (ibid.).

By 1990, Rutagarama had returned to Rwanda to work as a government park biologist at the Dian Fossey Gorilla Research Center in Volcano National Park. As part of his research into mountain gorillas and their habitat, Rutagarama conducted a study involving the rate of bamboo regeneration in the Virungas, marking his study areas with fluorescent sticks. In October 1990, however, a Tutsi-led rebel army invaded from Uganda and fought against government soldiers in the park. The mostly Hutu government soldiers used the rebel attack as an excuse to round up and imprison Tutsis—often for no reason other than that they were educated or had traveled abroad. The soldiers arrested Rutagarama, claiming that he had placed the fluorescent sticks in the jungle to guide Tutsi rebels. Rutagarama spent four months in prison before Tutsi rebels invaded the town of Ruhengeri in northern Rwanda and freed him in early 1991. He escaped to Burundi, where he was

reunited with his wife and children, who had escaped Rwanda a short time earlier.

In 1994 the ongoing ethnic conflict in Rwanda escalated into genocide, in which nearly 1 million people were killed—most of them Tutsis. Rutagarama was out of the country at the time, but he lost his parents and three brothers to the violence. Hundreds of thousands of Rwandan refugees fled through the Virunga mountains to the DRC. These desperate people poached animals for food and cut trees for shelter and fuelwood. Some retreating rebel forces also laid landmines and set booby traps throughout Volcano National Park.

Rutagarama returned to Rwanda later in 1994 to accept an assignment from the country's new government to take over the national park system. The civil war had destroyed the infrastructure of the parks, and the genocide had taken the lives of many rangers and other staff members. Furthermore, the huge number of refugees needing to be repatriated to Rwanda strained the resources of the government and put the continued existence of the nation's parks in jeopardy. Many of these refugees lived in or near the Virungas, and their settlements threatened to destroy mountain gorilla habitat—or the animals themselves through the transmission of human diseases.

Rutagarama created a strategic plan to rehabilitate the Office Rwandais du Tourisme et des Parcs Nationaux

(continues)

(Rwandan Office of Tourism and National Parks, or ORTPN). His plan, for which he garnered both public and governmental support, ensured that the protected areas retained their status and that gorilla habitat was not overrun by refugees. Part of Rutagarama's plan involved recruiting both Hutu and Tutsi park rangers and making them work together. "I think my behavior is the reason people in the park service trusted me—because I never asked their ethnicity," he stated. "I gained their confidence because I didn't have a hatred attitude" (ibid.).

In 1997, Rutagarama switched his focus from national to regional conservation efforts to save the mountain gorilla and its habitat. He became program director for the International Gorilla Conservation Program (IGCP)—an organization funded jointly by the African Wildlife Foundation, Fauna and Flora International, and the World Wide Fund for Nature. In this position, Rutagarama oversees gorilla conservation activities in Rwanda, Uganda, and the DRC. His efforts have resulted in joint patrols of the Virunga mountain region by rangers from all three countries. The three governments also share information and conservation strategies, despite the fact that Rwanda is at war with the DRC. In fact, Rutagarama has repeatedly risked his life to deliver food and equipment to rangers in the DRC, some of whom have not been paid in five years.

In 2001, Rutagarama received the prestigious Goldman Environmental Prize for his work in protecting the mountain gorilla from the effects of the Rwandan genocide as well as more recent conflicts in the DRC. He planned to use some of his $125,000 in prize money to help local communities near the parks and to attract tourists to his country. He hoped that conservation efforts might one day unite his country behind a common goal, for the benefit of both gorillas and humans. "After a humanitarian disaster as horrific as genocide, the common struggle to preserve something of shared value, like the natural environment, can form an ideal for people to believe in," he stated. "The opportunity and obligation to protect something precious can assist the reconstruction of a devastated society" (Goldman 2001).

Sources:

"Goldman Prize Recipient Profile: Eugene Rutagarama." *Goldman Environmental Prize*. 2001. Available at http://www.goldmanprize.org/recipients/recipientProfile.cfm?recipientID=108 (accessed February 28, 2003).

Rutagarama, Eugene. 2001. "A Conservation Triumph: The Mountain Gorillas of Rwanda." *Science in Africa*. Available at http://www.scienceinafrica.co.za/2001/July/gorilla.htm (accessed February 28, 2003).

Snell, Marilyn Berlin. 2001. "Gorillas in the Crossfire." *Sierra* 86 (November–December).

tourism sector. Around the world, growing wealth and leisure time, decreasing travel costs, and increased interest in outdoor activities have all combined to make tourism one of the fastest growing and most financially lucrative of industries. Indeed, the World Travel and Tourism Council has estimated that tourism provides employment for 10 percent of the global workforce, and the World Bank estimated the value of international tourism at the close of the 1990s at nearly U.S.$450 billion.

For some time now, the fastest-growing niche of tourism has been nature-based tourism. This genre of travel may account for as much as half of all tourism dollars, and it is increasing at a much swifter pace than other forms of tourism. But nature-based tourism is a generic, somewhat amorphous term that means different things to different people. For example, many conservationists draw clear distinctions between nature-based tourism and "ecotourism," an increasingly popular sobriquet that is not wholly understood or used appropriately. Whereas nature-based tourism merely describes the setting in which tourism activity takes place, true ecotourism has been defined by the Ecotourism Society as "responsible travel to natural areas that conserves the environment and sustains the well-being of local people." The World Conservation Union-IUCN, meanwhile, describes the ecotourism concept as "environmentally responsible travel and visitation to relatively undisturbed natural areas, to enjoy and appreciate nature (and any accompanying cultural features—both past and present) that promotes conservation, has low visitor impacts, and provides for beneficially active socioeconomic involvement of local populations." In these and other definitions, ecotourism—also sometimes dubbed sustainable tourism—"is distinct from 'nature,' 'adventure,' 'wildlife,' and virtually all other types of tourism because it focuses not simply on the type of leisure activity, but on tourism's impact and the responsibilities of both the tourist and those in the tourism industry (such as tour operators or lodge owners)" (Honey 1999b).

The Benefits and Pitfalls of Ecotourism

In many developing regions of the world, ecotourism concepts are centerpieces of national economic development efforts and conservation strategies. This is especially true in Africa, where countries including Kenya, Senegal, Namibia, Madagascar, Uganda, and Zimbabwe are all investing heavily in ecotourism programs. These efforts are being supported by a wide assortment of environmental nongovernmental organizations (NGOs) that see ecotourism as a viable conservation tool. Indeed, well-known groups such as the World Conservation Union-IUCN, Nature Conservancy, Audubon Society, Conservation International, Africa Wildlife Foundation, Sierra Club, and World Wide Fund for Nature have all become involved in ecotourism programs, studies, and field projects (Honey 1999a).

Advocates of ecotourism assert that when it is done correctly, it provides an avenue to bankroll the preservation of unique ecosystems while also providing economic opportunities for communities living near parks, reserves, and other protected areas. "Ecotourism can benefit protected areas through: exposure of the public to the natural world, with opportunities for improved environmental education and awareness, and consequently increased public support; generation of revenue, with the potential for this to be channeled back into protected area maintenance and management; and the creation of jobs in the region and the promotion of economic development, particularly for local communities" (Seddon 2000).

But many travel experiences that are touted as ecotourism actually fall far short of that ideal. For example, some alleged ecotourism travel packages display little interest in respecting the cultures or aiding the economies of rural or indigenous communities—an essential element of the ecotourism ideal. "Much of what the big players in the tourism industry sell as green tourism is known as 'ecotourism lite'—minor environmentally friendly, cost-saving measures (such as not washing sheets and towels each day) or 'add-ons' (a half-day hike into a rainforest or bird watching, for instance) to conventional vacations" (Honey 1999b).

In some of these cases, tour operators and tourism officials use the ecotourism label not out of any attempt to mislead, but out of a mistaken belief that the trips and destinations qualify as genuine ecotourism. Increasingly, however, private interests and governments alike are cynically utilizing the term as a money-making promotional tool, applying it to trips and destinations that provide only minor conservation or economic benefits to targeted regions.

Of course, even ecotourism that contributes to conservation goals and provides for the economic and political betterment of local residents can degrade destinations over time. "Protected areas are inherently sensitive sites; increased visitor levels will have an inevitable impact at a number of levels," observed one expert. "These negative impacts can lead to environmental degradation, economic inequity, and sociological change. . . . Negative impacts on protected areas will be exacerbated where the park or reserve lacks funds, lacks staff, lacks expertise, and is therefore unable to harness benefits for the protected area or for local communities. The message is simple: protected areas must specifically plan for ecotourism" (Seddon 2000).

Making Ecotourism Work in Africa

Africa was one of the first places in which early forms of ecotourism took root. Kenya, for example, was a global pioneer in applying ecotourism principles to its parks and reserves. Kenya and other East African countries such as Tanzania were also among the first to recognize that the future of threatened species and ecosystems is closely intertwined with the fortunes of human communities within and adjacent to protected areas. For example, poaching of monkeys, lions, elephants, and other creatures was far more likely among people that saw wildlife exclusively as a source of nutrition or crop-destroying pests, instead of as a potential vehicle for economic improvement. This epiphany led conservationists operating in East Africa in the late 1960s and 1970s to embrace a "stakeholders" theory of conservation that emphasized the importance of channeling economic benefits of wildlife and ecosystem preservation to area villages (Gakahu and Goode 1992; Roe 1997).

Today, African ecotourism takes a host of forms. In Zimbabwe, authorities have operated a Communal Area Management Programme for Indigenous Resources (CAMPFIRE) for the better part of two decades. Some of the operations within this program have appreciably improved quality of life for poor rural communities while simultaneously improving wildlife numbers. "Under CAMPFIRE, proprietary rights over wildlife have been devolved to communal area authorities where villagers manage local resources, including making contractual arrangements with safari operators for the lease of hunting and nonhunting tourism concessions. Key species such as elephant, buffalo, lion, and leopard provide the bulk of income from sport hunting. Because of their high value, they are tolerated and, consequently, conserved" (Getz 1999). Not surprisingly, these programs have been most effective in areas that are rich in wildlife and of limited suitability for farming, logging, and other activities.

Tanzania's Wildlife Division has implemented similar community wildlife management programs that are designed to be self-perpetuating by generating revenue for conservation and economic development from hunting safaris. "Tanzania's community-based sport hunting presents an ultimate paradox for ecotourism," acknowledged one expert. "While reprehensible to most wildlife lovers and ecotourism enthusiasts, hunting safaris can be lucrative—a hunter in Tanzania is estimated to spend 50 to 100 times more than a wildlife viewing tourist—as well as lower impact—hunters usually travel on foot, carry their supplies, and pitch tents, in sharp contrast to the hoards of minivans and luxury lodges that accommodate photo safaris" (Honey 1999b).

To date, many Tanzanian ecotourism initiatives are pale imitations of the ecotourism ideal. Many hunting-based projects remain deeply flawed, riddled by instances of mismanagement and disrespect for wildlife and ecosystems (as evidenced by episodes of excessive and inappropriate wildlife killing sprees). Tanzania's national parks system, meanwhile, now retains more of the revenue it generates from visitation. This has enabled Tanzania National Parks (TANAPA) to divert a modest amount of money for community development projects such as classrooms and water and sanitation systems. Most of the money generated by the parks, however, continues to go into the state treasury instead of into conservation or rural development programs. Still, Tanzania does boast a number of ecotourism success stories. For example, it harbors a number of private-sector ecotourism projects that are exemplary, both in terms of their environmental performance and their treatment of local communities (Honey 1999a).

Neighboring Kenya has a mixed record of fulfilling ecotourism ideals as well. Some parks and reserves pursue their wildlife and ecosystem conservation mandate with vigor, and a number of protected areas employ local residents as rangers, guides, and hotel staff and divert money generated by entrance fees and hotel operations into community projects. But corrupt officials have siphoned off some of this revenue, taking a toll on community development and park-community relations. Indeed, some park developments designed to accommodate—or take advantage of—rising visitation levels have been approved with little regard for community well-being or needs. These range from opening restaurants or hotels that make no provisions for local employment to converting natural areas valued by indigenous peoples for economic, cultural, or spiritual reasons.

Kenya's private wildlife reserves also have a mixed record. Some of these ranches, which are predominantly owned and operated by "white settler families who market an elegant but colonialist 'Out of Africa' experience under the banner of ecotourism" (Honey 1999b), have a track record of genuine environmental commitment. For example, some are involved in breeding endangered species such as the black rhinoceros or Rothchild's giraffe, while others are active members of the Ecotourism Society of Kenya, the continent's first such organization intended to set standards and promote ecotourism principles and practices. But some of these enterprises are loath to share the revenue they generate with nearby communities, and it has been charged that "these private reserves are an attempt to maintain family wealth and a lifestyle from a bygone era 'under the guise of conservation and ecotourism,' says Maasai activist Meitamei Ole Dapash" (ibid.).

It should be noted, however, that even flawed ecotourism ventures have a smaller environmental impact than other forms of tourism. They usually increase respect and appreciation for the natural world and its myriad creatures, contribute to conservation initiatives, and have a lower impact than conventional forms of tourism.

Realizing the Ecotourism Ideal

African countries seeking to improve their performance in the realm of ecotourism have been urged to take a number of steps. For example, wildlife and ecosystem conservation efforts would receive a tremendous boost if profits derived from ecotourism went directly to environmental protection, research, and education efforts associated with the destination park. One suggested measure to increase the size of these profits has been to increase park entry fees so that they can cover the park's capital costs and operating costs. Moreover, implementation of a multi-tier fee program would ensure that the bulk of revenue is garnered from wealthy foreign visitors rather than residents of modest economic means. In addition, increased use of special fees and tourism-based trust funds could channel tourist dollars directly to conservation (Vanasselt 2000).

African states also need to take additional steps to ensure that ecotourism initiatives are worthwhile for communities within and adjacent to parks and reserves. This can be accomplished by setting aside a portion of protected area revenue for local development projects or by emphasizing community involvement in the daily life of the protected area. For example, ecotourism planners "advocate sales of local handicrafts in gift stores, patronage of local lodges, use of locally grown food in restaurants and lodges, and training programs to enable residents to fill positions as tour guides, hotel managers, and park rangers" (ibid.).

African states also need to reduce corruption, bureaucratic wrangling, crime, and political unrest within their boundaries. These problems not only distract governmental attention away from environmental conservation and community development, but also contribute to declining visitation levels from foreign tourists. Finally, Africa's rural communities and its wildlife will benefit enormously if tourists genuinely committed to ecotourism ideals take it upon themselves to become more discriminating and knowledgeable in considering the legitimacy of trips and destinations marketed under the "ecotourism" label.

"Some experts have pronounced ecotourism dead, passé, or hopelessly diluted," acknowledged one expert. "However, amid the superficiality, hype, and marketing, there are excellent examples around the world of dedicated people, vibrant grassroots movements and struggles, and much creativity and experimentation. Although real ecotourism is indeed rare and usually imperfect, it is still in its infancy, not on its deathbed" (Honey 1999b).

Sources:

Africa Biodiversity Collaborative Group. 2002. *Impact of HIV/AIDS on the Management and Conservation of Natural Resources in Eastern and Southern Africa.* Washington, DC: ABCG.

Central African Regional Program for the Environment (CARPE). 2001. *Congo Basin Information Series: Results and Lessons Learned from the First Phase (1996–2000).* Washington, DC: Biodiversity Support Program.

Chenje, M., ed. 2000. *State of the Environment: Zambezi Basin.* Maseru/Lusaka/Harare: SADC/IUCN/ZRA/SARDC.

Friends of the Earth Middle East. 1999. *Protecting Open Spaces in the Middle East: Final Report on Symposium 11–13 July 1999.* FOE Middle East.

Gakahu, C. G., and B. E. Goode. 1992. *Ecotourism and Sustainable Development in Kenya.* Proceedings of the Kenya Ecotourism Workshop, Lake Nakurn National Park, Kenya, 13–17 September 1992. Nairobi: Wildlife Conservation International.

Getz, Wayne M., et al. 1999. "Sustaining Natural and Human Capital: Villagers and Scientists." *Science* 283 (March 19).

Greene, Michael J. B., and James Paine. 1997. *State of the World's Protected Areas at the End of the Twentieth Century.* Cambridge, UK: World Conservation Monitoring Center.

Honey, Martha S. 1999a. *Ecotourism and Sustainable Development: Who Owns Paradise?* Washington, DC: Island Press.

———. 1999b. "Treading Lightly? Ecotourism's Impact on the Environment." *Environment* 41 (June).

Hulme, D., and M. Murphree, eds. 2001. *African Wildlife and Livelihoods: The Promise and Performance of Community Conservation.* Oxford: Heinemann.

Llewellyn, Othman Abd-Ar-Rahman. 2000. "The WCPA Regional Action Plan and Project Proposal for North Africa and the Middle East." *Parks* (February).

McCullum. 2000. *Biodiversity of Indigenous Forests and Woodlands in Southern Africa.* Maseru/Harare: SADC/IUCN/SARDC.

McNeely, J. A., J. Harrison, and P. Dingwall, eds. 1994. *Protecting Nature: Regional Reviews of Protected Areas.* Gland, Switzerland: IUCN.

Muster, C. J. M., H. J. de Graaf, and W. J. ter Keurs. 2000. "Can Protected Areas Be Expanded in Africa?" *Science* (March 10).

Oteng-Yeboah, A. A. 1996. "Biodiversity in Three Traditional Groves in the Guinea Savanna." In L. J. G. van der Maesen et al., eds., *The Biodiversity of African Plants.* Dordrecht: Kluwer Academic.

Pearce, Fred, and Seamus Murphy. 2001. "Caught in the Middle." *Geographical* (December).

RAMSAR. 2003. *List of Wetlands of International Importance.* Gland, Switzerland: Bureau of the Convention on Wetlands. Available at http://www.ramsar.org/sitelist.doc (accessed May 2003).

Roe, N., N. Leader-Williams, and B. Dala-Clayton. 1997. *Take Only Photographs, Leave Only Footprints: The Environmental Impacts of Wildlife Tourism.* London: International Institute for Environment and Development.

Seddon, Philip J. 2000. "Trends in Saudi Arabia: Increasing Community Involvement and a Potential Role for Eco-tourism." *PARKS* 10 (February).

Shomon, Joseph James. 1998. *Wild Edens: Africa's Premier Game Parks and Their Endangered Wildlife.* College Station: Texas A&M University.

Stoddard, Ed. 2002. "South Africa Wages War on Water-Sucking Invaders." *Reuters News Service.* Available at http://www.planetark.org/avantgo/dailynewsstory.cfm?newsid=16742.

Sulayem, Mohammed S. A. "North Africa and Middle East." 2002. World Commission on Protected Areas (WCPA) Available at http://www.iucn.org/themes/wcpa/region/mideast/mideast.html.

Sweeting, J. E. N., A. G. Bruner, and A. B. Rosenfeld. 1999. "The Green Host Effect: An Integrated Approach to Sustainable Tourism and Resort Development." Washington, DC: Conservation International.

UN Educational, Scientific and Cultural Organization, Man and the Biosphere (MAB) Programme. 2002. "MAB Program World List of Biosphere Reserves." Available at http://www.unesco.org/mab/brlistAfr.htm (accessed February 2003).

UN Environment Programme. 2002. *Africa Environmental Outlook.* Nairobi, Kenya: UNEP.

Vanasselt, Wendy. 2000. "Ecotourism and Conservation: Are They Compatible?" In *World Resources 2000–2001.* Washington, DC: World Resources Institute.

Wells, M. 1997. *Economic Perspectives on Nature Tourism, Conservation and Development.* Washington, DC: World Bank.

4

Forests

Forests have long played an integral role in the lives of human communities throughout Africa and, to a much lesser degree, the Middle East. But the forests in these regions have experienced significant declines in volume and quality over the past half-century, a period during which unsustainable logging, agriculture, and consumption practices became entrenched in many societies. Today, these problems of deforestation and environmental degradation constitute a significant peril to the livelihoods of millions of people in the region, and they are endangering ecosystems across the continent.

The Benefits of Healthy Forests

Although the Middle East and Northern Africa feature only sparse forestlands, the importance of forests to wide swaths of Africa (and other regions of the world) can hardly be overstated. Human populations have relied on the forests for millennia to provide them with the raw materials for shelter, clothing, agricultural implements, and other tools. Moreover, woodlands supply plants used in traditional medicines and a wide variety of foods, including fruits, nuts, edible roots, honey, and protein-rich bushmeat (primarily small mammals and invertebrates). In addition, wood gathered from the forest is the primary fuel for cooking and heating in many parts of the continent. Forests and their by-products also are recognized as an important economic asset for many developing countries. Finally, forests and woodlands are deeply interwoven into the cultural, spiritual, and religious lives of many African peoples.

Forests also provide a host of direct and indirect environmental benefits. First and foremost, they nourish much of the continent's wealth of flora and fauna. Indeed, the broad green belt of tropical forest that girds Africa's mid-section—one of fourteen distinct forest types on the continent—contains an estimated 1.5 million species (World Conservation Monitoring Centre 2000). In addition, forests regulate precipitation runoff into rivers and streams, stabilize soil against erosion, and even influence climate by sequestering carbon and recycling rainfall as part of larger hydrological systems (Biodiversity Support Program 1992; Matthews et al. 2000).

But Africa's poor record of forest stewardship—which has been exacerbated in many regions by unforgiving climatic factors such as low precipitation and high temperatures—leaves the future of the continent's remaining forests and dependent flora and fauna in doubt. High rates of deforestation are a concern in nearly every corner of Africa, and efforts to address the primary anthropogenic factors driving this liquidation of ecological capital—unsustainable agricultural and logging practices, overgrazing, urban expansion, land tenure inequities, severe management funding shortfalls, and official corruption—remain inadequate in most countries.

Leading Causes of Forest Loss and Degradation

Timber and other resource extraction companies operating in Africa—including numerous foreign-owned firms—are a leading cause of deforestation and forest degradation on the continent. These industries are a significant presence for a number of reasons. First, some geographic regions are repositories of globally significant supplies of commercially valuable materials. Africa's rain forests contain several species of trees that are prized for the quality and appearance of their wood, and the continent's mineral wealth is breathtaking; indeed, West Africa has major deposits of gold, tin, and iron ore, and central and southern Africa contain a "mother lode" of industrial and precious diamonds, copper, and gold (Veit et al. 1995). In addition, some regions of Africa—most notably the continent's western reaches—contain significant oil and gas reserves. Moreover, regulation and monitoring of extractive activities is meager across much of the continent, in part because many African nations have sought to improve their economies by encouraging exploitation of their natural resources, and in part because of the funding shortfalls with which most of the countries perennially grapple. These factors provide extractive industries with little incentive to operate in environmentally sustainable ways, and have also been cited in the rise of illegal logging and mining activities across the continent (UN Environment Programme 2002).

Another leading cause of forest loss in Africa is escalating demand for land for farming and settlement, both of which are directly attributable to overall population growth. Indeed, in many areas of the continent, the scramble to secure basic necessities of survival such as food and shelter has trumped environmental considerations, leading to widespread livestock overgrazing, inap-

Pulp logging near Calabar in southern Nigeria. COREL

propriate agricultural practices, and infiltration into previously undisturbed natural areas. In Uganda, agricultural encroachment in Mount Elgon National Park destroyed more than 25,000 hectares of old-growth forest between 1970 and 1980. Since that time, the same scenario has played out over and over again in protected and unprotected forests across the continent. In Zimbabwe, for instance, some 700 square kilometers of forest are cleared each year just to grow tobacco (Chenje and Johnson 1994).

Use of forest resources for fuelwood has also been cited as a factor in forest loss, although the ultimate impact of this type of harvesting is in some dispute. Across much of Africa, rural industries consume substantial amounts of fuelwood to burn bricks, smoke fish, cure tobacco, brew beer, and dry salt. In countries such as Malawi, Mozambique, Tanzania, and Zimbabwe, the equivalent of one hectare of woodland (50 to 60 cubic meters of solid wood) is cut to cure each hectare of tobacco, and it has been reported that tobacco estates in Malawi clear about 100 square kilometers of woodland annually and account for about one-quarter of all fuelwood used in the country (Chenje and Johnson 1994; Bernard 1990). In addition, rural families use fuelwood culled from forests and woodlands to cook food and heat their homes. But some experts believe that these fuelwood users are quite discriminating in their consumption patterns. "Nearly all fuelwood used in rural households comes from indigenous trees, but household fuelwood collection did not, in past, contribute significantly to woodland clearance. It is often wrongly assumed that fuelwood collection is extensive and unselective. . . . In fact, rural people are highly selective about the fuelwood they collect. Certain types of trees and shrubs are trimmed or cut for fuel, particularly hardwoods which give a hot fire, make long-lasting coals, and produce little smoke. . . . Attempts to reduce deforestation with fuel-saving stoves have met with little success in southern Africa, primarily because the major cause of deforestation is not fuelwood use" (Chenje and Johnson 1994).

Another contributor to forest loss and degradation in Africa has been the ownership arrangement found in most countries. Across most of the continent, forests are state-owned, even though few countries have the financial, legal, and administrative frameworks in place to ensure that forest resources are managed responsibly and sustainably. This state of affairs not only leaves forests intensely vulnerable to destructive natural forces (such as fire, disease, and insect infestation) and anthropogenic forces (such as logging, mining, farming, and grazing) for agriculture and settlement, but also gives citizens little incentive to take the roles of environmental stewards when using the forest and its resources. "Secure tenurial arrangements and a clear understanding of roles and responsibilities are of central importance [to future African forestry policies]," stated the UN Food and Agriculture Organization. "Tenurial rights entail secure, long-term access, enabled by policies that recognize these local rights while also providing the holders with legal and regulatory support in protecting them—against the forest industry, state agencies and encroachment by other population groups, for example. Communities have to perceive that they will gain economically if their long-term commitment is to be secured. The incentives would be far greater if they received more productive forests, rather than degraded ones to manage. While economic benefits are critical, other benefits are also important. In the Gambia, it has been noted that, while the prospect of having relatively unhindered access to the forests and its benefits is important, the driving force behind community involvement has been the realization of the environmental benefits of forests and the satisfaction derived from the feeling of ownership" (UN Food and Agriculture Organization 2001a).

Regional Trends

Analyzing the current state of forest ecosystems in Africa and the Middle East is a formidable challenge. Many forest inventories in this region of the world are outdated, obsolete, fragmented, or of otherwise dubious value, and countries have historically adopted divergent definitions of "forest" when compiling data on forest cover, deforestation, afforestation, and other elements of

Table 4.1 Africa: Forest Resources by Subregion

Subregion	Land area	Forest area 2000		Total forest			Area change 1990–2000 (total forest)		Volume and above-ground biomass (total forest)	
		Natural forest	Forest plantation							
	000 ha	000 ha	000 ha	000 ha	%	ha/capita	000 ha/ year	%	m³/ha	t/ha
Central Africa	403,298	227,377	634	228,011	56.5	2.6	-852	-0.4	127	194
East Africa	590,078	134,132	1,291	135,423	23.0	0.7	-1,357	-1.0	28	38
North Africa	601,265	4,569	1,693	6,262	1.0	n.s.	33	0.5	32	51
Southern Africa	649,213	192,253	2,601	194,854	30.0	1.6	-1,741	-0.9	42	72
West Africa	733,359	83,369	1,710	85,079	11.6	0.4	-1,351	-1.5	61	84
Africa – small islands	1,181	130	107	237	20.1	0.1	4	1.9	88	121
Total Africa	2,978,394	641,830	8,036	649,866	21.8	0.8	-5,262	-0.8	72	109
TOTAL WORLD	13,063,900	3,682,722	186,733	3,869,455	29.6	0.6	-9,391	-0.2	100	109

SOURCE: UN Food and Agriculture Organization. 2000. *Global Forest Resource Assessment 2000*. http://www.fao.org/DOCREP/004/Y1997E/y1997e0g.htm#bm16.1. Accessed July 22, 2003.

forest health. By utilizing national experts, satellite technology, and other innovations, however, organizations such as the UN Food and Agriculture Organization (*Forest Resources Assessment 2000*) and World Resources Institute (*Pilot Analysis of Global Ecosystems: Forest Ecosystems*) have compiled detailed assessments of forest cover trends in Africa, the Middle East, and around the globe.

According to FAO estimations, Africa's total forest cover in 2000 was just under 650 million hectares, equivalent to 17 percent of the world total and 22 percent of Africa's total land area. Almost all of Africa's forest cover is located in its vast tropical ecological realm; indeed, the continent accounts for about one-quarter of the world's tropical rain forests. Silviculture, on the other hand, remains a modest factor in the continent's total forest area, as forest plantations accounted for only 1 percent of total African forests at the close of the twentieth century (UN Food and Agriculture Organization 2001a).

Africa contained about 0.85 hectare of forest per capita at the close of the 1990s, which is near the world average. But its per capita share of forest had been much higher just one decade earlier, when the continent held another 50 million hectares of forest. Indeed, Africa's annual net loss of forest area is now the highest among the world's regions, at –5.3 million hectares annually. This represents an average annual deforestation rate of nearly 0.8 percent between 1990 and 2000 (ibid.).

Not surprisingly, this rapid rate of deforestation has diminished continental stores of wild forests that still exist in large, intact ecosystems. These woodlands, which provide vital sanctuary for a startling array of flora and fauna, have been torn asunder in many regions of Africa. In West Africa, for instance, nearly 90 percent of the original forest is gone, and remnants are heavily degraded and fragmented. Only in Côte d'Ivoire and the border country between Cameroon and Nigeria can isolated patches of undisturbed forest be found. Similarly, Madagascar's tropical forests are justly famous for nourishing a spectacular range of animals and plants not found on the African mainland (or anywhere else). But these once-pristine forests—and their associated ecosystems—have been disrupted, fragmented, and degraded to the point that the island no longer contains any tracts large or natural enough to qualify as what the World Resources Institute describes as "frontier" forest (Bryant 1997).

Middle East (West Asia)

Forests are only a minor component of the natural landscape in the Middle East, a region of the world characterized by low rainfall, hot temperatures, and large expanses of desert. Those forests that do exist consist primarily of open woodlands with scattered trees, although the highlands of Turkey, Georgia,

Figure 4.1 Regional Comparison of Change in Forest Area, 1980–1995

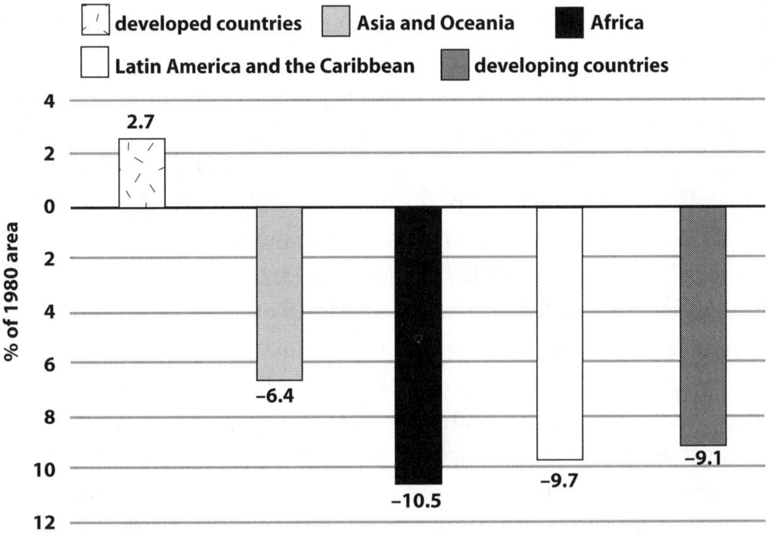

SOURCE: UN Environment Programme. 2002. *Africa Environment Outlook: Past, Present and Future Perspectives.* http://www.unep.org/aeo/023.htm. Accessed July 22, 2003.

Armenia, Azerbaijan, and Afghanistan support areas of closed-canopy forest. All told, the Middle East has less than 1 percent of the world's total forest cover, and only 3.2 percent of its total land area is classified as forest (UN Food and Agriculture Organization 2001a).

Turkey's forest resources are the greatest among the six Middle East nations that currently harbor more than 1 million hectares of forest land. Indeed, Turkey alone accounts for almost 38 percent of the region's total forest area. These natural forests are recognized as an important element of the country's overall economic security; according to one study, more than 17,000 villages located in or near Turkish forests depend on those forests for their livelihood (Duzgun and Ozu-Urlu 2000). At the other end of the spectrum, the only existing forests in Bahrain, Kuwait, Oman, Qatar, and the United Arab Emirates are plantations (UN Food and Agriculture Organization 2001a). These operations, which have also sprouted in Turkey, Iran, and elsewhere, emphasize fast-growing and multipurpose exotic tree species such as eucalyptus, poplar, and acacia that can be utilized for a variety of purposes, including industrial harvesting, fuelwood, charcoal production, and shelterbelts/windbreaks (Heywood 1997).

As in other areas of the world, leading sources of pressure on the Middle East's modest forest resources include agricultural expansion, overgrazing, urbanization, and increased demand for forest products, all of which are strongly linked to population growth. In addition, many rural populations continue to exploit forests for fuelwood and charcoal for domestic use. In 1998, for example, 98 percent of Lebanon's total consumption of roundwood was for fuel; other nations heavily reliant on fuelwood as an energy source include Iraq (67 percent of total consumption), Jordan (66 percent), Turkey (44 percent), and Saudi Arabia (41 percent) (UN Food and Agriculture Organization 2001a). Economic troubles have also hindered adoption of effective forest conservation measures, and wars have wreaked significant damage on forests in areas such as Iraq, Lebanon, and Afghanistan.

In recent years, several Middle East nations have taken steps to institute sustainable management practices in their stewardship of natural forests. Turkey, Cyprus, Syria, and Lebanon have all initiated national forestry programs, and there has been increased recognition of the importance of integrating forestry programs with agricultural and rural development institutions. In addition, several countries have actively sought to secure greater community involvement in managing and protecting forest resources, though most forests in the region remain state-owned (UN Food and Agriculture Organization, 2001b).

North Africa

The dominant landscape feature of this region of the African continent is the Sahara Desert, covering 6 million square kilometers. The shifting dunes of the world's largest desert cover 94 percent of North Africa's total land area, which in turn accounts for roughly one-fifth of the continent. Forests, by contrast, cover only 1 percent of the region's land surface and account for only 0.16 percent of the world forest area. This was not always the case, however. "In the past, under the combined effects of severe climate, growing populations and lack of adequate land use planning, the forest cover [in North Africa] suffered from large-scale deforestation," reported the UN Food and Agriculture Organization. "Clearing of forests and the use of fire for cultivation and grazing reduced the cover to patchy relics as compared to the reported cover present during previous centuries" (ibid.). Today, overgrazing, man-made fires, and drought conditions remain serious impediments to afforestation and conservation of natural forest remnants. Indeed, desertification has made a slow, methodical march across the region, making forest recovery an exceedingly difficult challenge.

As in other places in Africa, national forest inventories in this region are fundamentally flawed, riddled with obsolete, incomplete, or misleading data. National experts estimate, however, that Morocco, Algeria, and Tunisia account for more than 90 percent of the forest cover in North Africa, even though they have less than 50 percent of the region's total land area. The presence of forests in these countries is directly attributable to the comparatively high levels of annual precipitation that they receive in their northern ramparts (ibid.). But many of these forests are wilting under the pressure of human activities. In Morocco, for instance, the total area of natural forest is in steady decline and the natural balance for many forest ecosystems "has been irreversibly disrupted" (Benabiud 1996). The leading causes of forest loss— estimated at 30,000 hectares annually in the mid-1990s—include overgrazing, unsustainable logging, land clearing for farming and urban expansion, and fuelwood gathering. Authorities have approved extensive reforestation programs as a means of counteracting these losses of natural forest, but success rates have been low, in part because efforts are undertaken with an emphasis on short-term economic gain—by planting exotic species vulnerable to climate, insects, or diseases, for example—rather than natural forest regeneration. "As Morocco's human and livestock populations continue to increase, the pressure they exert on forests will become ever more onerous. The condition of Morocco's forests is now reaching a crisis stage. Both from environmental and socioeconomic perspectives, urgent measures are needed to halt further deterioration of these forests" (ibid.).

Morocco aside, trends in forest cover area are moving upward in the region. In Egypt, where the amount of forest is so small that any reforestation is significant, forest cover has been growing at a 3.3 percent annual clip. Libya (1.4 percent) and Algeria (1.3 percent) have also reported heartening increases in forest cover because of tree planting efforts and new resource conservation policies—though again, base forest areas in these countries are very small. By the close of the 1990s it was estimated that planted forests accounted for 27 percent of North Africa's total forest area.

Forest management plans are also being introduced on an incremental basis across much of the region. In Algeria, where more than 91 percent of forests are state-owned, 28 percent of the country's total forest area was covered by a formal management plan at the close of the twentieth century. Moreover, Algeria has specifically targeted its remaining natural forests when adding to its system of protected areas (Ikermoud 2000). These kinds of efforts, which are also evident in Tunisia and Morocco, have helped to minimize desertification in the northern parts of those countries (UN Food and Agriculture Organization 2001a).

West Africa

Forest resources in West Africa are modest, at approximately 11 percent of the total regional land area, 13 percent of the continent's total forest cover, and 2 percent of the world forest area. This sparse level of forest cover is attributable to arid conditions in the northern part of the region (Mauritania, Mali, and Niger), high population densities (Togo, Niger, Benin, and Nigeria), and deforestation at the hands of agricultural and logging interests (Côte d'Ivoire). Mauritania and Niger have the smallest forested areas (0.3 and 1.0 percent of their total land area, respectively), while Guinea-Bissau is the region's most forested country, with 60 percent of its land area classified as forest (ibid.).

Within this region, the more northern countries of Chad, Mali, Mauritania, and Niger account for 65 percent of the land area in this subregion but feature forest cover on only 6 percent of their combined territory. Indeed, the forests of West Africa are almost exclusively located in the southern part of this region, in belts of tropical rain forests and dry forest located to the south of the vast Sahara Desert. Cote d'Ivoire, for instance, is blessed with more than 225 species of trees in its forests, including obeche, mahogany, and iroko. But these and other regional forests have suffered considerable damage in recent years, and the long-term outlook for them is uncertain. West Africa's annual rate of forest loss (−1.5 percent on average) is almost twice that of the continent as a whole (−0.78 percent), and several countries are going through their forests at an alarming rate. In terms of area, Nigeria and Cote d'Ivoire have posted the greatest annual losses of forest cover in recent years. But Niger has the highest annual deforestation rate, losing 3.7 percent annually during the 1990s (ibid.).

The chief culprits in the loss and degradation of West African forests include rapid population growth and widespread poverty, both of which spur short-term exploitation of forest resources for fuelwood, food, building materials, and land. Indeed, fuelwood is believed to account for as much as 85 percent of the total energy consumption in this region, and demand has been so heavy that periodic shortages have been reported in Benin, Nigeria, Togo, and elsewhere in recent years (Bellefontaine et al. 2000). Other factors include environmentally unsound economic development, reliance on shifting cultivation and other destructive farming practices, industrial mining and logging, and government policies that alienate rural communities and indigenous peoples that might otherwise be valuable allies in sustainable forest management. Illegal logging in forests, which remain overwhelmingly state-owned, is also a significant issue because regional governments lack the financial resources or political will to enforce restrictions or to negotiate beneficial concessionaire agreements (UN Food and Agriculture Organization 2001a). Timber exploitation in Cote d'Ivoire, for example, has been characterized as "sheer anarchy" by the International Institute for Environment and Development. Finally, violent conflicts in places like Liberia and Sierra Leone have pulverized some forested areas and led to the settlement of large refugee communities in other previously undisturbed woodlands.

Liberian Timber Trade Blamed for Bankrolling Regional Violence

The forests of Liberia are an important piece of the Upper Guinean Forest Ecosystem, a band of rain forest that has been cited as one of the world's treasures of biodiversity. It is believed to house some 10,300 species of plants and vertebrate animals, including the last remaining viable populations of the pygmy hippopotamus and Liberian mongoose. But this forest is also one of the most threatened natural areas on the globe (Mittermeier et al. 1999). The leading culprit in Liberia's rapid forest loss is industrial logging, which is taking place at a breathtaking pace. According to investigations undertaken by an array of organizations, from the United Nations to international NGOs (nongovernmental organizations) like Global Witness and Greenpeace, the revenue gained from these timber sales is being funneled to paramilitary and rebel groups that have been blamed for much of the violence and civil unrest that has plagued the region in recent years.

Analysts believe that the administration of Liberian president Charles Taylor has long used the country's natural wealth to fuel bloodshed in West Africa. In the 1990s, the primary natural resource used to buy weapons and expand armies was diamonds. Money generated by the export of diamonds was subsequently distributed to Liberian government forces—including its notorious Armed Terrorist Unit (ATU)—to crush armed resistance of the rebel group LURD (Liberians United for the Return of Democracy), as well as to help the rebel Revolutionary United Front (RUF), which, since 1991, has been engaged in a vicious struggle to overthrow the government in neighboring Sierra Leone.

By 2000 the increasingly explicit link between Liberia's diamond sales and West Africa's brutal civil wars had sparked international condemnation, and in the spring of 2001 the United Nations formally agreed to a global embargo on so-called conflict diamonds exported by Liberia. But Taylor anticipated this development. At his urging, the Liberian parliament adopted the Strategic Commodities Act of 2000. This law gave Taylor de facto control over much of the country's natural resources, from iron ore and diamonds to timber. Since that time, evidence suggests that Liberia now relies on sales from timber exports to bankroll both the RUF in Sierra Leone and its own domestic military operations. In fact, critics charge that Liberia is dramatically understating its actual volume of timber cut and value of revenues from timber sales. According to some estimates, the annual income generated by Liberian timber exports rose from U.S.$5 million to $U.S.100 million from 1995 to 2000 (Global Witness and International Transport Workers Federation 2001), making it the single largest source of export earnings for the Liberian government. In addition, the logging roads that have been carved through the region have emerged as significant

(continues)

pipelines for the delivery of weapons from Liberia to RUF rebels in Sierra Leone.

The ecological toll from this accelerated program of forest liquidation has been further exacerbated by the nonexistent regulatory environment in which loggers operate. Indeed, the timber companies that are slogging their way through Liberia's forests are "free of even the most cursory environmental and humanitarian controls" (Brackenbury 2002). The potential consequences of this situation are considerable, ranging from terrorization and displacement of rural communities (in some cases by logging companies that have built their own private militias) to extirpation of endangered species or severe flood events, siltation of rivers and streams, and other changes associated with deforestation.

In December 2000 and October 2001, two separate UN advisory panels recommended a UN embargo on Liberian round log exports. But attempts to institute such a ban were blocked by France and China, which together account for two-thirds of all imports of Liberian wood. In the meantime, Taylor's government granted a logging concession of 16,000 kilometers—more than 40 percent of Liberia's remaining productive forest— to the Oriental Timber Company (OTC), a firm that has been suspected of repeated forays into illicit arms and drugs trafficking (ibid.).

Sources:

Brackenbury, Andrew. 2002. "Liberian Logs Fuel War." *Geographical* 74 (August).

Global Witness and International Transport Workers Federation. 2001. *Taylor Made: The Pivotal Role of Liberia's Forests and Flag of Convenience in Regional Conflict.* September. Available at www.oneworld.org/globalwitness/liberia/liberia_report.htm.

Greenpeace International. 2002. "Liberian Timber Trade Fuels Regional Insecurity." Greenpeace report available at http://archive.greenpeace.org/forests/forests_new/html/content/reports/liberia.pdf (accessed March 2003).

Mittermeier, Russell A., Norman Myers, and Cristina Goettsch Mittermeier. 1999. *Hotspots: Earth's Biologically Richest and Most Endangered Terrestrial Ecoregions.* Mexico City: CEMEX, Conservation International.

UN Food and Agriculture Organization. 2001. *Global Forest Resources Assessment 2000.* Rome: FAO.

All of these factors have contributed to the desertification crisis gripping the nations within and bordering the southern Sahara. "Considerable effort has been expended to stop and even reverse this trend, including reforestation with exotic species, green belt plantations and agroforestry development," noted one analysis. "There has also been a great deal of scientific research on nitrogen-fixing trees, which are important in the conservation of soil fertility. The utilization of these trees has led to spectacular results in sand dune fixation in Senegal" (ibid.). But the strategy of establishing plantations to halt or reverse desertification has been only fitfully effective. Moreover, the rise of monocultural timber plantations has not kept pace with the rate of exploitation of natural forests, which support complex ecosystems that plantations cannot possibly duplicate (UN Food and Agriculture Organization 2000a).

Despite these grim trends, however, several nations are taking encouraging steps to protect their dwindling forest resources. Most West African countries are instituting sustainable management concepts into new forestry policies, and awareness of the need to secure local community cooperation in meeting conservation goals is growing. Integration of forest conservation principles into agricultural and rural development institutions is also increasing. Experts warn, however, that such programs will ultimately be for naught if the current rate of forest exploitation—both legal and illegal—is not meaningfully addressed.

Central Africa

Central Africa is home to the continent's largest and most biologically diverse forests. As with other regions of Africa, national forest inventory data in this area is fragmented, outdated, or nonexistent. According to expert analysis, however, Central African forests constitute 35 percent of Africa's total forest area and about 6 percent of the globe's total forest cover. These forests—which cover approximately 57 percent of Central Africa's total land area—constitute the second-largest contiguous expanse of tropical rain forest in the world, next to the Amazon. The rain forest encompasses Gabon, Equatorial Guinea, the Congo, large portions of Cameroon and the Democratic Republic of the Congo, and a piece of the Central African Republic, and it accounts for nearly two-thirds of the African continent's total forest biomass (UN Food and Agriculture Organization 2001a).

Gabon is the most forested country in Central Africa, with 85 percent of its total area blanketed by tropical tree species that support a wealth of flora and fauna (Burundi has the lowest proportion of forest cover in the region, at 4 percent of total land area). But in terms of sheer volume, the Democratic Republic of the Congo (DRC), formerly known as Zaire, is dominant; it contains more than 60 percent of the subregion's forest area. Not surprisingly, the nation's forests are also a treasure trove of biodiversity and endemism. More than 1,100 species of birds and 400 species of mammals roam the forests of

A logger cuts down a tree in the Gabon rainforest. GALLO IMAGES/CORBIS

the DRC, which also contain an estimated 11,000 plant species, 30 percent of which are endemic (found nowhere else in the world). In addition, these dense forests house many of the continent's remaining tribes of pygmy people, who depend on the forest for shelter, medicine, food, and the continuation of their cultural traditions (ibid.; Bryant 1997).

To date, reforestation programs in Central Africa have been modest in scope. In addition, the emphasis has been on creation of new commercial plantations rather than reforestation of areas degraded by logging, mining, or farming activity. Many of the region's plantations, which have an uneven record of success, are located in Burundi and Rwanda, both of which launched extensive plantation programs from the late 1970s through the early 1990s (UN Food and Agriculture Organization 2000b).

Historically, the forests of Central Africa have not been subject to the intense logging pressure that prevails in other areas of the continent. In the Democratic Republic of the Congo, for instance, the dense, old-growth forests of the vast Congo Basin have remained intact in large part because the nation's stunted transportation system has precluded large-scale timber and mineral exploitation. In fact, some remote areas of the DRC have fewer passable roads and smaller human populations than in 1960, when the country gained independence from Belgium (Bryant 1997). Commercially attractive tree species also tend to be widely scattered in the forest, which relieves pressure from industrial logging operations (UN Food and Agriculture Organization 2001a). But the region inherited poor forest management services from France (several countries are former colonies of the French), and economic pressure on the forests is strong.

Indeed, degradation and outright clearing of natural forests is believed to be escalating in the DRC, as well as in other forest-rich countries like Cameroon and the Congo. In Gabon, meanwhile, a similar scenario seemed to be unfolding, with about three-quarters of its forests logged or allocated for logging concessions. But in 2002 the country delighted the conservation community by announcing that it was setting aside approximately 10 percent of its total land area—including large tracts of vulnerable forest—in a newly created national park system.

Loggers in the region typically concentrate on high-value tree species, leaving other species untouched, and the logging roads that are left behind have opened up the forest to bushmeat hunters who have ravaged some animal populations, to prospective farmers who practice unsustainable brands of agriculture, and to refugees from civil unrest in neighboring countries. All three of these intertwined groups have taken a noticeable toll on the ecological integrity of forests throughout the Congo Basin (Bryant 1997). During the 1990s, for example, the forests of Burundi and Rwanda suffered devastating damage when desperate refugees from regional war zones cleared large tracts for agricultural use and fuelwood.

The vexing, persistent problem of illegal logging has also garnered increased attention in recent years. In Cameroon, for instance, conservationists have estimated that half of the 2 million hectares of rain forest that were harvested between 1980 and 1995 were taken by illegal logging operations. But institutionalized corruption has made it difficult to halt these rogue timber operators, whose offenses include cutting trees in unauthorized or protected areas, exceeding permitted volumes, falsification of documents, and nonpayment of taxes. For example, in 1999 the government formally banned the export of timber cut from its forests and passed regulations meant to ensure that the processing remained in Cameroon. But in the eastern part of the country, where large tracts of forest are controlled by Europe-based timber companies, there is only one monitor for every 20,000 hectares of tenured forest. "If [the monitors] report infractions, sometimes the paperwork 'disappears' in exchange for bribes paid to more senior officials or reports are sometimes held in file for officials to extort money from companies that would otherwise be fined. These monitors also often find themselves working alone in an area where the logging company may be the largest employer. They are often threatened with violence or open to offers of bribes" (Elliott 2002).

Besides logging, the chief factors driving forest loss and degradation include commercial, residential, and industrial development pressure (especially in Gabon and the Congo, which are heavily urbanized), agricultural expansion (most notably in the DRC, Burundi, Cameroon, and Rwanda), and poor management of natural resources at the national and regional levels (Central African Regional Program for the Environment 1996). Another factor is Africa's heavy consumption of fuelwood, though there has been considerable debate about its effect on forest health. Indeed, the cumulative impact of fuelwood collection on Central Africa's forests is difficult to gauge because of the dearth of quantifiable data on the activity. However, it is widely acknowledged that wood from the forest remains the primary source of domestic energy for most communities in the region.

Many nations in Central Africa grant that they need to improve their stewardship of their forests, which rank among the region's greatest natural assets. But Gabon excepted, the countries of Central Africa are among the poorest in the world, and political instability and violent civil wars have characterized much of the region's recent history. As a result, financial, political, technical, and management institutions are ill equipped to implement effective protection strategies for remaining forests, which are overwhelmingly state owned. Indeed, many forestry administrations lack basic resources to fulfill their duties effectively. For example, the last national forest inventories conducted in Burundi and the Democratic Republic of the Congo took place more than two decades ago, in 1976 and 1982, respectively. This absence of current and accurate information on the state of forest resources makes it exceedingly difficult for institutions to pursue conservation goals in an efficient manner (UN Food and Agriculture Organization 2001a; Central African Regional Program for the Environment 1996).

Unfortunately, the dysfunctional, calcified forest management systems that typify the region are likely to persist for some time to come. In Cameroon, for instance, forestry officials still do not require loggers to file forest management plans, and they render cutting limitations moot by allowing subcontracting of harvest permits (Elliott 2002). But conservationists are heartened by some tangible signs of change. In Gabon, for example, resource inventories and forest management plan proposals must be submitted before any form of exploitation is permitted, and pilot projects for sustainable forest production have been implemented in the Central African Republic and Cameroon (UN Food and Agriculture Organization 2001a).

A number of Central African countries have also designated large forest areas as parks or other types of protected areas. These are sensible, farsighted steps, for heavily forested national parks like Dzanga-Ndoki National Park in the Central African Republic and Dja National Park in Cameroon support a terrific array of biodiversity. But there is significant room for improvement in managing these parks, and in expanding existing protected area systems (ibid.; Fotso 1996).

East Africa

East Africa has little reliable national data on its forest resources, but international experts believe that the subregion accounts for approximately 21 percent of the continent's total forest area and about 4 percent of the world's forests. Nearly all of this area's forest resources are natural forests; of its total forest area of 135.4 million hectares, only about 1.3 million hectares are located on plantations. In terms of individual countries, Sudan has the largest percentage of its land under forest cover (46 percent), followed by Tanzania (29 percent) and Kenya (13 percent) (UN Food and Agriculture Organization 2001a).

Nightmarish living conditions are prevalent in several countries in this region of the world. Shelter, clothing, food, and potable water have long been prized here. But the quest for these basic necessities of life has become much more hazardous and difficult in recent decades because of civil wars and armed conflicts between neighboring nations. This violence, which has wracked Sudan, Ethiopia, Eritrea, Somalia, and Rwanda in recent years, has reduced some countries to virtually lawless states and produced huge refugee populations that have, in turn, taken an enormous toll on some regional ecosystems.

Deforestation is proceeding at a dizzying pace in several countries. Uganda has the region's highest annual rate of deforestation (2 percent)—in part because of the presence of a large refugee population that is reliant on forest land for building materials, agricultural fields, and fuelwood. Indeed, by some estimates, forests supply about 90 percent of Uganda's total energy demand (ibid.). But in terms of sheer volume of trees lost, Sudan is the leader. It is estimated that the troubled nation is losing nearly 1 million hectares to deforestation on an annual basis. This loss of woodlands, which provide valuable soil retention, moisture catchment, and windbreak functions, have exacerbated the country's losing battle with desertification. Indeed, the UN Convention for Combating Desertification has declared that fully half of Sudan's twenty-six states are feeling the effects of desertification. Other countries grappling with the specter of desertification include Eritrea, Ethiopia, Djibouti, and Somalia (ibid.).

Another factor in the plummeting fortunes of regional forests is forest fires. In Ethiopia, where the amount of land area covered by forest has declined from 40 percent to 3 percent over the space of 40 years (1960–2000), fires set by farmers as a land-clearing tool have caused serious economic and environmental damage in recent years. In 2000, for example, immense wildfires, the result of land-clearing activities, blackened the skies over the southwestern forests of Bale and Borena for three months, despite the efforts of 15,000 firefighting personnel. By the time the flames were extinguished, more than 300,000 hectares of natural forest had been lost or damaged, and the total cost of the Bale and Borena fires was estimated at $U.S.39 million. Yet according to some observers, even these enormously destructive fires have not spurred Ethiopian authorities to "give adequate attention to efficiently protecting its last natural forest resources" (Lemessa 2002; UN Food and Agriculture Organization 2001b).

Analysts agree that afforestation, reforestation, and fire prevention programs are required across much of East Africa to counteract the destructive activities (livestock grazing, shifting cultivation, land-clearing by fire) and drought conditions that are withering the region's dwindling forest resources. They cite recent efforts in Eritrea, where modest programs have been launched to replant indigenous tree species, nurture new plantations, and protect fragile natural areas with significant ecologic value from development or exploitation, as steps in the right direction (UN Food and Agriculture Organization 1997). Increasing community involvement in conserving forest resources, which are primarily state-owned, has also been cited as a priority. "Land tenure is perhaps the single most important factor in natural resources management, environmental degradation, and fire use," concluded one analysis of Ethiopia's forest policies. "Without changing ownership either literally or symbolically to give local communities a greater sense of investment in the land, environmental disasters will continue and the 2.7 percent of the country [of Ethiopia] that is forested will rapidly diminish" (Lemessa and Perault 2002).

Southern Africa

The size, health, and ecological importance of forests vary considerably among the nations in the southernmost reaches of the continent. In places like Zambia, Angola, Madagascar, and Mozambique that typically receive moderate or generous amounts of precipitation, natural forests remain an important part of ecosystems and serve as an important stimulus for economic growth. By contrast, the arid conditions that prevail in countries like Namibia, Botswana, South Africa, and Lesotho have hindered natural forest growth, and decades of exploitation and agricultural conversion have reduced the forests of Zimbabwe and Malawi to small fragments of their former selves (UN Food and Agriculture Organization 2001a).

African Forests Devastated by Fires Set by Farmers

Forest fires are a major cause of forest loss in Africa, and many of these blazes have been attributed to farmers who use fire as a tool in land-clearing efforts. Indeed, in many regions of Africa, deliberately set fires have been a staple of agricultural practice. But while most fires remain under control, those that break containment often cause enormous havoc given the dry climate conditions and limited forest resources that prevail across much of the continent. In Ethiopia, for instance, forest fires blamed on land-clearing activities destroyed more than 300,000 hectares of natural forest in 2000.

Historically, Africa's developing countries have experienced little success in instituting fire prevention and fire suppression programs. A primary reason for this woeful track record is the lack of financial resources and institutional support for such endeavors. But observers have also pointed to the forest ownership dynamic that is in place in most countries as an underappreciated factor in forest loss from fire. Forests and other natural resources are almost exclusively owned and operated by African states. This situation reduces incentive for local communities to practice environmentally sustainable methods of exploitation. But evidence suggests that community attitudes can undergo significant change if governments grant ownership or long-term user rights to forest resources.

Gambia, for example, introduced community forestry on a pilot basis in 1991, began implementing it on a countrywide scale in 1994, and in 1998 fully endorsed the transfer of forest ownership to rural communities. As these steps were taken, newly empowered communities exhibited a striking new level of effectiveness in preventing and fighting wildfires. According to one study, the occurrence of fires declined dramatically in the area of the country in which community forestry was started. The study further indicated that community management produced gains in enforcement of other environmental laws related to forest health (UN Food and Agriculture Organization 2001).

Sources:

Lawry, Steven W. 1990. "Tenure Policy toward Common Property Natural Resources in Sub-Saharan Africa." *Natural Resources Journal* 30, no. 2.

Makombe, Kudzai, ed. 1994. *Sharing the Land: Wildlife, People and Development in Africa*. Harare: IUCN.

UN Food and Agriculture Organization. 2001. *State of the World's Forests 2001*. Rome: FAO.

Veit, Peter G., Adolfo Mascarenhas, and Okyeame Ampadu-Agyei. 1995. *Lessons from the Ground Up: African Development that Works*. Washington, DC: World Resources Institute.

At the close of the twentieth century, southern Africa's forest cover accounted for about 5 percent of the global total (its total land area is also roughly 5 percent of the world total). But it accounts for a disproportionate share (30 percent) of Africa's forest cover, since its total land area is less than one-quarter (22 percent) of the continental total. Three nations account for the lion's share of forest area: Angola contains 36 percent of the subregion's forested land, while Mozambique and Zambia each hold 16 percent. Combined, they account for approximately 68 percent of the subregion's forest cover, even though they have less than 45 percent of the land area (ibid.).

In most nations of the region, forest cover is in a state of decline, both in terms of quantity and quality. But rates of forest loss and degradation hinge on a host of factors, including precipitation levels and other climatic conditions, population and economic growth, agricultural trends, and conservation policies. Some nations, including Angola and Mozambique, have managed to keep their deforestation rates at less than 0.2 percent annually (UN Food and Agriculture Organization 2001b). But analysts note that many natural forests in these countries have escaped exploitation only because of internal turmoil. Civil wars rendered many rural areas unsafe, and prompted large-scale flight to population centers. But since 1992, when peace was restored in Mozambique, forests have come under increased pressure from farmers, loggers, and other people eager to rebuild their lives (UN Food and Agriculture Organization 2001a).

The highest deforestation rates in southern Africa are in Malawi and Zambia, where pressure to convert forests for cultivation and settlement is high. In fact, Zambia alone has been losing about 850,000 hectares annually in recent years, which accounts for 49 percent of the total deforestation in the entire subregion forests (Zambia Ministry of Environment and Natural Resources 1998; UN Food and Agriculture Organization 1999). As in other parts of Africa, rural communities are leading consumers of forest resources in Zambia. In these communities, where poverty is a constant companion and famine a looming threat, families are dependent on forests and associated flora and fauna for energy, shelter, food, employment, and trade. But government-sanctioned exploitation has rarely been conducted with an eye toward sustainability, either. "In countries where timber is exploited under licenses for selected species and quantities without regard to proper silviculture, the forests have been deeply degraded or even stripped of a number of their most valuable species and their biodiversity adversely affected," commented the UN's Food and Agriculture Organization. "The loss of forest cover is contributing to soil erosion, causing water pollution and siltation of rivers and dams [in southern Africa]" (UN Food and Agriculture Organization 2001a).

In order to address the region's historic lack of natural forest cover—as well as its more recent patterns of unsustainable forest exploitation and consumption—countries have pursued a strategy that relies heavily on creation and maintenance of plantations. Indeed, nations such as South Africa and Swaziland have brought significant numbers of productive, healthy forest plantations into being. Swaziland's Usutu pine plantation is the largest single plantation on the continent, at more than 700 square kilometers, and managed woodlots account for 9.3 percent of its total land area. South Africa, meanwhile, accounts for 60 percent of the region's plantations and nearly 20 percent of the entire continent's plantation acreage (UN Food and Agriculture Organization 2001b).

But some of these operations have been established with insufficient regard for the welfare of indigenous species or ecosystem preservation. In many places, native forests have been replaced with commercial woodlots of eucalyptus, pine, and other exotic species. In some instances, these exotics have thrived to the point that they have escaped plantation boundaries and compromised nearby indigenous forests. Elsewhere, the comparatively greater thirst of exotic species for water has damaged some fragile freshwater ecosystems. In South Africa, for instance, streamflows on Cathedral Peak were reduced by almost half within two decades of a major introduction of pines in the watershed. Overall, tree plantations have reduced river runoff by 2.4 percent in South Africa, an arid country that reverberates with every change in water availability (Huntley 1991).

In addition, many indigenous plants and animals cannot survive in monocultural plantations composed of exotic tree species, for the latter cannot duplicate the complex ecosystems found in natural forests (Chenje and Johnson 1994). For example, blue swallows disappeared from Zimbabwe's Eastern Highlands in the 1970s after its natural habitat was sacrificed to make way for new plantations (Huntley 1991). Similarly, many of the rare and endangered plants in South Africa's Transvaal region are located in areas that have been targeted for tree plantations.

Sources:

African Development Bank. 2001. *Gender, Poverty and Environmental Indicators on African Countries 2001–2002.* Adbijan: African Development Bank.

Albert, Jeff, et al. 1998. *Transformations of Middle Eastern Natural Environments: Legacies and Lessons.* New Haven, CT: Yale University Press.

Alden Wily, L. A., and S. Mbaya. 2001. *Land, People and Forests in Eastern and Southern Africa at the Beginning of the 21st Century.* Nairobi: World Conservation Union.

Bellefontaine, R., A. Gaston, and Y. Petrucci. 2000. *Management of Natural Forests of Dry Tropical Zones.* FAO Conservation Guide No. 32. Rome: FAO.

Benabiud, Abdelmalek. 1996. "Forest Degradation in Morocco." In Will D. Swearingen and Abdellatif Bencherifa, eds., *The North African Environment at Risk*. Boulder, CO: Westview.

Bernard, M. P. 1990. "Improved Efficiency of Biomass Usage in Malawi." In *Proceedings of the First World Renewable Energy Congress*. New York: Pergamon.

Biodiversity Support Program. 1992. *Central Africa: Global Climate Change and Development—Synopsis*. Biodiversity Support Program (BSP)—World Wide Fund for Nature, The Nature Conservancy and World Resources Institute. Landover, MD: Corporate Press.

Bojang, F. 2000. "Overview of the International Workshop on Community Forestry in Africa." In *Proceedings of the International Workshop on Community Forestry in Africa*. Rome: FAO.

Bryant, Dirk, D. Nielson, and L. Tangley. 1997. *The Last Frontier Forests: Ecosystems and Economies on the Edge*. Washington, DC: World Resources Institute.

Central African Regional Program for the Environment. 1996. *CARPE Workshop, Libreville, Gabon*. Washington, DC: USAID.

Chenje, Munyaradzi, and Phyllis Johnson, eds. 1994. *State of the Environment in Southern Africa*. Harare, Zimbabwe: South African Research and Documentation Centre, IUCN-World Conservation Union, and Southern African Development Community.

Duzgun, M., and E. Ozu-Urlu. 2000. *Forest and Forestry Policy Development in Turkey: Country Report*. Cairo: FAO.

Elliott, Valerie. 2002. "Rainforests Are Falling to Greed and Corruption." *Times Online*, April 17. Available at http://www.timesonline.co.uk.

Erdmann, Thomas K. 1992. *An Analysis of Ten African Natural Resource Management Practices*. Washington, DC: Forestry Support Program.

Fotso, C. 1996. *Problematique de la conservation de la biodiversité en Afrique centrale*. Libreville, Gabon: USAID.

Fuggle, R. F., and M. A. Rabie, eds. 1992. *Environmental Management in South Africa*. Kenwyn: Juta and Co.

Global Forest Watch. 2000. *A First Look at Logging in Gabon*. Washington, DC: World Resources Institute.

Hegazy, A. K. 1999. "Deserts of the Middle East." In M. A. Mares, ed., *Encyclopedia of Deserts*. Norman: University of Oklahoma Press.

Heywood, H. 1997. *The International Expert Meeting: Plant Resources and Their Diversity in the Near East, 19–21 May 1997*. Cairo: FAO Regional Office for the Near East.

Huntley, B. J., ed. 1991. *Biotic Diversity in Southern Africa*. Cape Town: Oxford University Press.

Ikermoud, M. 2000. *Evaluation des ressources forestières nationales*. Algiers, Algeria: Direction Générale des Forêts.

Lemessa, Dechassa, and Matthew Perault. 2002. *Forest Fires in Ethiopia*. Addis Ababa: UN Emergencies Unit for Ethiopia.

Makombe, Kudzai, ed. 1994. *Sharing the Land: Wildlife, People and Development in Africa*. Harare: IUCN.

Matthews, Emily, Richard Payne, Mark Rohweder, and Siobhan Murray. 2000. *Pilot Analysis of Global Ecosystems: Forest Ecosystems.* Washington, DC: World Resources Institute.

Obasi, Godwin O. P. 2002. "Embracing Sustainability Science: The Challenges for Africa." *Environment* 44 (May).

UN Environment Programme. 2002. *Africa Environment Outlook.* Hertfordshire, UK: Earthprint Limited and UNEP.

UN Food and Agriculture Organization. 1996. *Forestry Policies of Selected Countries in Africa.* Rome: FAO.

———. 1997. *Support to Forestry and Wildlife Subsector.* Rome: FAO.

———. 1998. *Overview and Opportunities for the Implementation of National Forest Programmes in the Near East.* Damascus, Syria: FAO.

———. 1999. *State of the World's Forests 1999.* Rome: FAO.

———. 2000a. *Actes de l'atelier sous-régional sur les statistiques forestières et perspectives pour le secteur forestier en Afrique/FOSA sous région ECOWAS.* Yamoussoukro, Côte d'Ivoire: FAO, Rome.

———. 2000b. *Collecte et analyse de données pour l'aménagement durable des forêts— joindre les efforts nationaux et internationaux.* EC-FAO Subregional Workshop for Congo Basin Countries, Lambarene, Gabon, 27 September–1 October 1999. FAO: Rome.

———. 2001a.*Global Forest Resources Assessment 2000.* Rome: FAO.

———. 2001b. *State of the World's Forests 2001.* Rome: FAO.

Veit, Peter G., Adolfo Mascarenhas, and Okyeame Ampadu-Agyei. 1995. *Lessons from the Ground Up: African Development That Works.* Washington, DC: World Resources Institute.

Wolfire, Deanna M., Jake Brunner, and Nigel Sizer. 1998. *Forests and the Democratic Republic of Congo: Opportunity in a Time of Crisis.* Washington, DC: World Resources Institute.

World Conservation Monitoring Centre. 2000. *Global Biodiversity: Earth's Living Resources in the 21st Century.* Cambridge, UK: World Conservation Press.

Zambia Ministry of Environment and Natural Resources. 1998. *Zambia Forestry Action Plan, Volume 1—Executive Summary.* Lusaka: MENR.

5

Agriculture

Agriculture is an integral part of the socioeconomic fabric of many African and Middle Eastern communities, from the Arabian Peninsula to South Africa's Cape of Good Hope. Many of these societies are subsistence family-oriented ventures, but commercial agricultural enterprises are also vital. Indeed, the latter type of farming is the single largest employer in many African nations, and commercial agriculture makes meaningful contributions to national economic growth and export earnings.

Unfortunately, the current state of agriculture in both Africa and the Middle East is poor. In sub-Saharan Africa, where dependence on rain-fed agriculture systems is high, production from crops and livestock falls far short of what is needed to feed rapidly growing human populations, given the region's limited financial capacity to import food. This gap between output and need is attributable to a number of related factors, including widespread impoverishment, high rates of infectious disease (such as AIDS), civil unrest, and government mismanagement. All of these elements have limited people's capacity to grow or purchase food. In 2002 an extended drought further aggravated these problems, creating famine conditions for an estimated 16 million Africans. This grim situation, which is merely the latest in numerous drought-famine events that have devastated the region over the past half-century, further underscores the basic inadequacy of existing economic and agricultural systems. In the Middle East, meanwhile, widespread implementation of irrigation systems has made the region's agricultural interests less vulnerable to the vagaries of weather. But by all accounts, consumption of extremely limited freshwater resources is unsustainable in all countries and across all sectors, from households to irrigation-dependent farmers. Moreover, Middle Eastern rangelands used by pastoralists for generations are in a serious state of decline, with mismanagement of water resources, overgrazing, and drought all contributing to high rates of desertification.

Agriculture in the
Middle East and North Africa

In terms of climate, the Middle East (also known as the Near East or West Asia) and North Africa are arid regions that receive very low and highly variable annual rainfall. This fundamental reality has historically had a major influence on crop selection, geographic concentrations of farming activity, and the types of agriculture practiced across the region's 11 million square kilometers (divided roughly evenly between the Middle East and North Africa) (World Bank 2002). Despite the limitations imposed by climate, however, agriculture was the mainstay of the regional economy as recently as a half-century ago. Since that time, though, economic development has focused on the industry sector—specifically the petroleum industry—and agriculture has become less important. At the close of the 1990s, agricultural products accounted for only 5 percent of exports from the Middle East and North Africa, on account of enormous volumes of oil shipped overseas to the United States and other destinations (ibid.). Still, subsistence forms of cultivation and pastoralism endure in many parts of the region, and agriculture continues to supply a significant portion (10 to 25 percent) of total gross domestic product (GDP) in such countries as Syria, Yemen, Egypt, and Turkey. For North Africa as a whole, the agriculture sector accounted for 13 percent of the region's GDP in 1999, and it employs 30 to 40 percent of the overall workforce (World Bank 2001).

Indeed, while oil revenue has eclipsed all other economic sectors in the region, many governments still see national food security as a priority. Policies reflective of this attitude have historically emphasized agricultural protectionism, the erection of trade barriers, and government subsidies for agricultural inputs ranging from irrigation water to pesticides and fertilizers. Critics contend, however, that many of these subsidies have been ultimately detrimental to the future of agriculture in the region because they have encouraged unsustainable or destructive uses of finite land and water resources.

For instance, national imperatives to realize higher levels of food security spurred a tremendous increase in total irrigated area from the early 1970s to the end of the 1990s across the Middle East, from less than 3 million hectares to nearly 7.2 million hectares (FAOSTAT 2001). But while this rise in production capacity fueled a 4 percent agricultural growth rate across the region from 1980 through the mid-1990s (World Bank 2002), the increase in food production has still not kept pace with population growth, and the manner in which these gains were realized has wreaked major undesired changes on the environment. Today, salinization, alkalization, water logging, and nutrient depletion attributable to inefficient irrigation and other forms of water mismanagement are apparent across much of the area. Salinization, which is the most important cause of degradation in irrigated soils, has affected about 42.5 percent of the desert area across the Middle East, and more than half of the irrigated lands in the Euphrates plains in Syria and Iraq have been significantly degraded by salinization and waterlogging (UN Economic and Social Commission for Western Asia 1997; UN Environment Programme 2002b).

Regional agricultural efforts have also been hurt in recent years by drought, despite the presence of irrigation infrastructure. From 1999 to 2001, many parts of the Middle East and North Africa experienced successive years of drought that resulted in contraction or stagnation of farming sectors. Of the various crop types, cereals have been the most severely harmed by the lack of rain. In North Africa, for example, cereal production dropped by almost 10 percent from 1999 to 2000 because of drought conditions (UN Food and Agriculture Organization 2002).

Pastoralists have also been hurt by the cumulative impact of the droughts. Livestock accounts for between 30 and 50 percent of total agricultural GDP in the Middle East/North Africa region, and livestock losses caused by malnutrition or disease have seriously imperiled the livelihoods of owners, who live primarily in remote areas where artificial water sources are scarce. In recent years, herders have watched helplessly as drought conditions have withered vital rangeland vegetation and reduced the availability of feed. Indeed, feed became so scarce—and hence expensive—that many herders of cattle, sheep, and other livestock sacrifice purchases of household staples in order to keep their animals fed.

All told, drought is estimated to have affected at least 40 percent of the region's livestock populations from 1999 to 2001, according to the UN Food and Agriculture Organization. Heavy losses resulting from animal mortality and production losses (in volume of dairy goods, for example) have been widely reported in most countries, and the decline in food availability has spurred widespread distress sales of livestock, saturating the market so that sellers can reap only a fraction of the livestock's normal value. For instance, the average purchase price for a live sheep dropped by almost 50 percent between 1999 and 2000 alone (ibid.).

The Future of Agriculture

Water scarcity, desertification, and other forms of land degradation loom as the major agricultural issues confronting the Middle East and North Africa for the twenty-first century. Governments need to take swift action in implementing more sustainable models of agrarian and pastoral activity; otherwise, desertification and water shortages will become even greater problems than they are today, and the region will experience further diminishment of its limited supply of cropland, rangeland, and intact natural areas (Council of Arab Ministers Responsible for the Environment 1996).

Rangeland currently accounts for about half of the total land area in the Middle East, but it has been estimated that 90 percent of this land has been degraded or made vulnerable to desertification by drought, overgrazing, and poor stewardship of water resources (UN Environment Programme 2002b). The latter two factors are directly attributable to human activities and policies. For example, studies indicate that the grazing intensity in most Middle East countries has more than doubled over the past four decades, mainly as a result of subsidized feeding, provision of water points, and mechanization. In some locales, the increase has been even more ecologically unsustainable. In the West Bank, for instance, it is estimated that the grazing capacity on rangelands is exceeded by a factor of 5.7 (Palestinian National Authority 2000).

The exhaustion of rangelands has serious consequences for the region's nomadic pastoralists, who are directly dependent on the rangelands' quality and quantity for their livelihoods. In years of normal rainfall, livestock are kept on the rangeland for eight months and then fed for the remaining four. But recent drought conditions and years of overgrazing have created acute shortages of forage and drinking water, forcing herders to shoulder the burden of purchasing feed and water for several additional weeks—or even months—each year. The financial repercussions of this reallocation of household income are being felt in rural communities throughout the region. "This phenomenon requires immediate attention to prevent major population displacements and further environmental degradation," warned the UN Food and Agriculture Organization. "This long period of drought has caused significant damage to the environment and to the region's biological diversity, including both animal and plant species. Wildlife [that competes with livestock] has been severely affected as a result of the shortage of drinking water, lack of feed, dried wetlands, and degradation of wildlife habitats" (UN Food and Agriculture Organization 2002). The introduction of "safety net" measures such as livestock insurance would help pastoralists, as would increased investment in health and education services for rural communities. But ultimately, recovery of rangelands will not occur if pressure on these fragile lands is not reduced.

In addition to relieving stress on overused rangelands, regional governments will have to make substantive changes in the way they use water for agriculture. Across much of North Africa and the Middle East, irrigation systems are taking too much water from both surface sources and aquifers. They are also using it inefficiently, creating serious problems with salinity and water logging and raising the distinct possibility of serious water shortages down the line, especially since continued population growth and predicted rainfall diminishment associated with climate change will further stretch extremely limited water resources. Yet current water policies in most countries of the Middle East and North Africa provide little or no financial incentive for water conservation. This will have to change quickly if nations with significant agriculture sectors hope to maintain viable operations in the second half of the twenty-first century.

Fortunately, some of the counterproductive policies and behavior that have exacerbated the region's unsustainable use of resources in the agriculture sector are being addressed. Individual action plans to combat desertification are proliferating across the Middle East and North Africa, with a special emphasis on passing new laws and regulations restricting water consumption and land use. In addition, most countries in the Middle East have united to establish a regional action plan to address desertification issues within the framework of the UN Convention to Combat Desertification (UNCCD). Other sustainability initiatives are being pursued as well, such as the creation of carefully managed "range reserves" that will operate under sustainable use guidelines. Finally, some nations are linking environmentally friendly agricultural practices to other policy areas. In the United Arab Emirates, for example, President Sheikh Zayed bin Sultan Al Nahyan's acclaimed "greenification" project addresses energy, freshwater, and sustainable agriculture issues in one fell swoop. The project utilizes solar power, urban and industrial wastewater, and modern drip and traditional Bedouin irrigation techniques in a wide array of agricultural and habitat restoration schemes. Thus far, the project has planted more than 40 million productive date palm trees of thirty-seven different varieties, replenished coastal mangrove habitats, expanded total forest cover by one-third, and created "green belts" around cities that afford protection from sandstorms while simultaneously providing habitat for wildlife (Salloum 2001).

Agriculture in Sub-Saharan Africa

Sub-Saharan Africa's total land area covers 24.3 million square kilometers, the majority of which is arid or semiarid (World Bank 2002). Within this territory, agriculture is the cornerstone of most African economies and the activity around which most African lives revolve. The livelihoods of an estimated 65 to 70 percent of sub-Saharan Africa's population center on crop cultivation or livestock care, and agriculture accounts for one-third of the continent's total gross domestic product (GDP) and about 40 percent of its total

Coffee farms in Africa account for about 16 percent of global production. COREL

export earnings (International Food Policy Research Institute 2002). Africa's major commercial crops include cocoa (Africa produced two-thirds of the world's cocoa in 2001), coffee (16 percent of the world total in 2001), cereals, cotton, fruit, nuts and seeds, oils, rubber, sugar, tea, tobacco, and vegetables (UN Food and Agriculture Organization 2002).

Large tracts of previously undeveloped land have been converted to agriculture across Africa over the past three decades, with most of the net gain in tillable fields rather than pastureland. By the close of the twentieth century, more than 200 million hectares of the continent were under cultivation (up from 166 million hectares in 1970), while another 900 million hectares were used as rangeland by pastoralists (UN Food and Agriculture Organization 2001). However, the percentage of agricultural land (cultivated and pasture) varies considerably across Africa. Nearly 55 percent of southern Africa's land is devoted to agriculture, as is almost 47 percent of the land in the Western Indian Ocean Islands. By contrast, less than 20 percent of land in Central Africa—home to the continent's major tropical forests—has been set aside for agricultural purposes, though development pressure on these species-rich forests is growing.

The expansion of farming into previously undisturbed woodlands and savanna has driven overall increases in production volume since the early 1970s,

Figure 5.1 Value of Production per Agricultural Worker, 1995–1997 Average

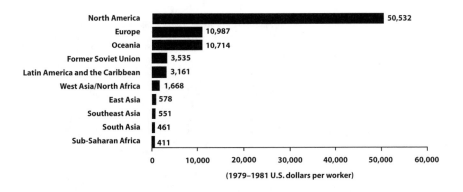

SOURCE: Reproduced with permission from the International Food Policy Research Institute. http://www.ifpri.org/pubs/books/page/agroeco.pdf. and World Resources Institute

though improvements in cultivation methods and increased inputs of fertilizers and pesticides have also played a part. For example, cereal production in Africa nearly doubled from 1975 to 1999 (UN Food and Agriculture Organization 2001). Yet these production volume gains mask major inefficiencies in the sector. Per capita output of staple foods have been trending downward in recent years, and the continent's world market share of cocoa, coffee, and other major export crops is eroding. In addition, these gains have not kept pace with the galloping rate of population growth in sub-Saharan Africa. In fact, the number of people suffering from malnutrition in Africa has doubled since the early 1970s (UN Food and Agriculture Organization 2002).

The role—if any—that economic reform programs have played in this surge in malnutrition rates has been the subject of considerable debate. Since the early 1980s, many sub-Saharan African governments have sought to reduce state involvement in the economy in favor of a more market-oriented approach, though advances have been halting. In the agricultural sector, this change in philosophy resulted in the elimination of some price controls on agricultural commodities, privatization of some state farms and other enterprises in the agriculture sector, reduction of taxes on agricultural exports, and reduction of subsidies for fertilizers. In some cases, these measures appear to have had their desired effect. In Kenya, for example, decentralization of the dairy industry and associated decontrol of milk pricing produced a major

Figure 5.2 Malnutrition Rates Are Falling—Except in Africa

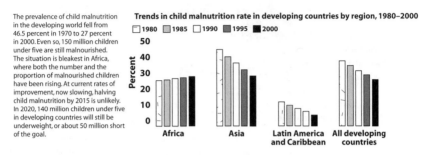

The prevalence of child malnutrition in the developing world fell from 46.5 percent in 1970 to 27 percent in 2000. Even so, 150 million children under five are still malnourished. The situation is bleakest in Africa, where both the number and the proportion of malnourished children have been rising. At current rates of improvement, now slowing, halving child malnutrition by 2015 is unlikely. In 2020, 140 million children under five in developing countries will still be underweight, or about 50 million short of the goal.

Trends in child malnutrition rate in developing countries by region, 1980–2000
☐ 1980 ▨ 1985 ☐ 1990 ▨ 1995 ■ 2000

Africa   Asia   Latin America and Caribbean   All developing countries

SOURCE: World Bank. 2002. *World Development Indicators 2002.* p. 40. http://www.worldbank.org/data/wdi2002/people.pdf. Accessed July 22, 2003.

surge in dairy production, with per capita production levels rising to twice that of anywhere else on the continent (Gabre-Madhin and Haggblade 2001). But the implementation of these reforms has drawn mixed reviews. "Proponents argue that the reforms have improved market efficiency, reduced budget deficits, stimulated export production, and increased the share of the final price received by farmers. Opponents argue that the reforms have destabilized agricultural prices, widened the income distribution gap, and reduced access to low-cost inputs" (Kherallah et al. 2002).

At the consumption end, meanwhile, winners and losers created by reform efforts vary from country to country. In nations such as Zambia and Zimbabwe, deregulation has forced poor urban families to pay more for maize, rice, and other staples. But in some urban centers in eastern and southern Africa, the urban poor have benefited from the newly competitive environment, which has produced lower marketing margins and lower food prices. In any case, state involvement in agriculture and other economic sectors—whether it takes the form of outright state ownership of plantations or subsidy programs—remains high in many countries, despite these recent initiatives (ibid.).

These reforms, and the aforementioned expansion of farming activity into previously undeveloped areas, have also failed to keep pace with the continent's swelling population. Today, sub-Saharan African population growth rates are among the highest in the world, even though warfare and disease claim millions of lives every year. Civil war and armed conflict between nations have ravaged many parts of the continent, and HIV/AIDS, malaria, and other infectious diseases have devastated countless families and communities.

Indeed, sub-Saharan Africa is believed to hold 70 percent of the adults and 80 percent of the children living with HIV/AIDS around the world (Hunter 2000). Populations are also increasing despite the persistence of crushing poverty and recurrent famine events. In fact, "Africa is the only continent where hunger and poverty are projected to worsen in the next decade. New and continuing crises appear likely to further disrupt agriculture, create refugees, escalate the need for and costs of emergency relief, and divert investment from the long-term solutions Africa so desperately needs to end its cycle of despair. Improving the poor performance of Africa's stagnating agricultural sector, in recent decades one of the worst in the world, is the key to solving the problems of hunger and poverty" (International Food Policy Research Institute 2002a).

Many African communities have frantically sought to address the pressing issue of food security by squeezing higher levels of productivity out of existing fields. But this practice has contributed to alarming increases in soil erosion and desertification. Other farmers are extending their activities onto previously untilled lands. But this strategy of extending cultivation and grazing often converts areas that are of only marginal agricultural value, creating a dynamic in which vital wildlife habitat and natural resources (such as river catchment areas) are destroyed, degraded, or altered in exchange for only modest

Table 5.1 Food and Nourishment in Africa (1996–1998)

Subregion (countries)	Food Availability	Prevalence of undernourishment		Depth of undernourishment
	Ave. per capita dietary energy supply (kcal/day)	Proportion of population undernourished (%)	Number of undernourished (millions)	Average food deficiency (kcal/person)
Northern Africa (6)	3,055	8*	10.7*	183
Central Africa (6)	1,898	50	38.5	344
Eastern Africa (7)	1,833	42	52.2	359
Southern Africa (9)	1,736	45	28.6	302
Western Africa (14)	2,570	16	33.0	238
Indian Ocean Countries (2)*	2,475	23*	3*	245

*NOTE: Data for the undernourished people in Northern Africa have doubled because of the poor situation in Sudan, and the same is applicable in the IOC with respect to Madagascar.

SOURCE: World Bank. 2002. *World Development Indicators 2002* p. 40.
http://www.worldbank.org/data/wdi2002/people.pdf. Accessed July 22, 2003.

gains. And increasing water scarcity across much of the continent precludes most states from introducing the modern irrigation measures that have boosted production in Asia and elsewhere. At the close of the twentieth century, less than 3 percent of the available cropland in sub-Saharan Africa was irrigated, yet the agriculture sector still accounted for 63 percent of all freshwater withdrawals in Africa in 1995, and numerous countries, including Burkina Faso, Chad, Egypt, Ethiopia, Gambia, Guinea, Libya, Madagascar, Malawi, Mali, Mauritania, Morocco, Mozambique, Niger, Rwanda, Senegal, Sierra Leone, Somalia, Sudan, and Swaziland, directed more than 80 percent of their freshwater use to the agriculture sector (Gleick 2000).

With increased irrigation an exceedingly limited option, sub-Saharan Africa remains dependent on rain-fed agriculture, the fortunes of which rise and fall with annual rainfall patterns. In years of drought, crops wither and famine stalks the land. Moreover, production from these rain-fed fields may become even more tenuous in the future, especially in arid regions like the Horn of Africa and the Sahel, if predicted changes associated with climate change—"global warming"—come to pass.

Today, rising populations are forcing many African nations to allocate an ever greater percentage of their limited funds to importing food staples, leaving less money to address pressing shortcomings in education, health, environmental protection, economic development, and agricultural modernization. This is exacerbated by stagnant or declining revenues from food exports, which have suffered from trade barriers in foreign markets, heavy agricultural subsidies in Western countries, and fluctuations in international food prices (Kherallah et al. 2002; Mosely 2003). Finally, many African states produce a narrow range of agricultural products, which increases their economic vulnerability to disease, pests, price fluctuations, and weather (UN Environment Programme 2002a).

Famine Conditions in Sub-Saharan Africa

Drought, grinding poverty, government mismanagement, war, and expanding populations have combined to create famine conditions in sub-Saharan Africa on several occasions in the last half-century. The most recent famines of regional scope occurred in 2000, when millions of people in at least sixteen African countries experienced food shortages either because of crop failure or because violence and warfare battered crop yields and disrupted distribution networks, and in 2002, when an estimated 16 million Africans once again were classified as being at risk of starvation.

The 2002 famine was initially centered in the southern African nations of Angola, Lesotho, Malawi, Mozambique, Swaziland, Zambia, and Zimbabwe.

Zimbabwe held the largest population at risk, with an estimated 7.2 million people in need of food. The totals for the other countries, according to the UN World Food Program, were: Malawi (3.3 million, primarily in the south), Zambia (2.9 million), Angola (1.8 million), Lesotho (650,000), Mozambique (600,000), and Swaziland (270,000). But across the continent, the total number of Africans at risk of starvation was placed at 40 million, because of emerging crises in the Horn of Africa (where there were 18 million in need of assistance, including 11 million in Ethiopia alone), Central Africa (2.7 million), West Africa (800,000), and the Western Sahel (450,000) (UN World Food Program 2003).

The crisis was created by the usual overriding factors—successive years of drought, compounded by disease, poverty, failed agricultural policies, civil unrest, and government mismanagement or corruption. But specific factors not seen before also came into play, including Zambia's decision to reject aid packages of genetically modified (GM) food because of concerns about the technology's long-term safety; Malawi's ill-conceived decision to sell all of its Strategic Grain Reserve—at a loss—in 2001; the AIDS pandemic, which struck down tens of thousands of villagers too weak to tend their fields; and

Sudan's Failed Bid to Become the Breadbasket of the Arab World

In the early 1970s, Sudanese leaders became convinced that a major program of investment in mechanized farming could transform the country into an agricultural powerhouse in the Arab world. This ambitious program was formally titled the Mechanized Farming Cooperation initiative, but it was more commonly known as the Habila scheme, after the fertile land that was the centerpiece of the program.

Sustainable forms of cultivation and pastoralism had long been practiced in the Habila region by the indigenous Nuba people. But when the Sudanese government imposed their mechanized farming plans in the 1970s and early 1980s, the subsistence lifestyle of the Nuba was dismantled. Nuba farmlands were seized without compensation, and their capacity to pursue pastoral activities was drastically curtailed by a variety of government edicts. Sudan then set up a leasing system for the land with the understanding that plots would be developed for mechanized rain-fed agriculture. The new system attracted wealthy Arab merchants who boosted crop volume significantly, but the scheme's emphasis on production resulted in a rapid decrease in the amount of fallow land and incursions into unsanctioned "illegal" land. By 1985, about 45 percent of mechanized agriculture was located outside

(continues)

sanctioned areas (UN Environment Programme 2002).

By the mid-1990s, the wealthy farmers who had purchased the leases on the Habila land had established a pattern of behavior in which they cultivated the land to exhaustion, then abandoned it in favor of fertile lands elsewhere, where they repeated the process. The Nuba farmers who had once tilled the land for themselves had meanwhile either been relegated to laboring anonymously in the fields of the absentee owners or pushed from the region entirely. Stripped of their traditional sources of livelihood and subject to violent raids from Sudanese troops battling rebel forces in the country's long civil war, more than half of all the Nuba people have scattered to other parts of the impoverished country (UN Office for the Coordination of Human Affairs 2002).

By the close of the twentieth century, extended periods of drought, civil war, and the poor land stewardship practices introduced by the mechanization scheme had made the Habila region far less productive than it had been twenty years earlier, when subsistence forms of cultivation and pastoralism had been practiced. In addition, the region's unfair tenure systems have contributed greatly to the political tensions and violence that have plagued the country for the past two decades.

Sources:

Shazali, Salah, and Abdel Ghaffar M. Ahmed. 1999. "Pastoral Land Tenure and Agricultural Expansion: Sudan and the Horn of Africa." Paper for the DFID Workshop on Land Rights and Sustainable Development in Sub-Saharan Africa, Berkshire, UK, 16h –19 February. Available at www.iied.org/pdf/dry_ip85.pdf (accessed February 2003).

UN Environment Programme. 2002. *Global Environment Outlook–3 (GEO-3)*. London: Earthscan.

UN Office for the Coordination of Human Affairs. 2002. *Sudan: Consolidated Inter-Agency Appeal 2002-Revision (March)*. New York and Geneva: United Nations.

the end of Angola's nearly three-decade-long civil war. The latter development marked the close of a terrible chapter in the nation's history, but it also triggered an influx of refugees back into the country, where they found that they had no means to cultivate their battered land or purchase basic food staples (Abrams 2002; UN World Food Program 2003).

Finally, the virtual collapse of Zimbabwe's agricultural operations from 2000 to 2002 had serious consequences for all of southern Africa, since the region has long relied on food pulled from Zimbabwe's fields for sustenance. In 2000 Zimbabwean president Robert Mugabe launched a controversial land-reform policy to appropriate farmland long held by white commercial farmers and give it to black citizens. The campaign was touted as an initiative to correct a shameful legacy of the country's colonial era. However, the subsequent program used violence and threats to redistribute the country's prime farmland, and Mugabe's forces showed little consideration for keeping the country's agricultural output stable during this transitional period. Instead, production levels plummeted as the appropriated land was handed over to supporters of the Mugabe government and white farmers departed the country, taking their farming expertise and investment capital with them. Citizens without connections to the Mugabe government—which includes most subsistence farmers—have received little or no land from this scheme (Abrams 2002; Robinson 2002).

In early 2003 the World Food Program stated that food shortages in Zimbabwe had worsened to the point that more than half of the country's population—about 7.2 million people—needed emergency food aid. But donor food is reaching only about 4.5 million of those at risk, and international agencies in Zimbabwe have expressed concern that the state is purposely directing food aid away from people deemed unfriendly to the government (World Food Program 2003).

Improving Agricultural Performance in Sub-Saharan Africa

Improving long-term agricultural productivity in sub-Saharan Africa is a formidable task, necessitating major revisions in land use, fundamental changes in prevailing farming and pastoral practices, increased state investment in rural health and education, changes in inequitable land tenure arrangements that have endured since the colonial era, and improvements in government and private sector performance, both within Africa and in the international arena. Effectively addressing these and other issues will not be easy, quick, or inexpensive. But if present trends continue unchecked, the continent faces a future of even greater misery and famine than it has endured in recent years.

Following are five major areas in which the nations of sub-Saharan Africa have been urged to make changes in the near future: environmental degradation of natural resources; inequitable land rights; increased investment in agriculture programs (including determination of the suitability of genetically modified crops); rural services and infrastructure; and improved institutional performance.

ENVIRONMENTAL DEGRADATION OF NATURAL RESOURCES

Rampant desertification from unsustainable farming practices, overgrazing, and deforestation—which are in turn manifestations of rapid population growth—is one of the most alarming environmental danger signs sweeping across the continent, as it is a central factor in Africa's growing inability to feed

This village near Kiffa in Mauritania has been decimated by desertification, a growing problem across Africa. AFP/CORBIS

its people. By the close of the twentieth century, signs of desertification—defined by the UN Convention to Combat Desertification as "land degradation in arid, semiarid, and dry subhumid areas brought about by climatic variations and human activity"—could be seen on 13 million square kilometers across Africa (43 percent of the continent's total land area), home to an estimated 270 million people (Obasi 2002; UN Environment Programme 2002a). According to one recent study, 55 percent of that vast area was at high or very high risk of desertification, which would be a crushing blow to countless rural communities (and numerous species of flora and fauna (Reich et al. 2001).

Fortunately, the UN Environment Programme states that Africa has taken a leadership role in engineering a global response to the desertification issue. It was essential in creating, adopting, and ratifying the UN's Convention to Combat Desertification (CCD), which in turn has prompted a number of African nations to pass "action plans" specifically designed to combat the desertification threat. Elements of these plans include public education initiatives, reforestation and other land enhancement activities, and improved environmental monitoring (UN Environment Programme 2002a).

Soil erosion is another problem that has been exacerbated over the years by unsustainable land use. Currently, high levels of soil erosion are silting up vital

Figure 5.3 Main Causes of Desertification by Region

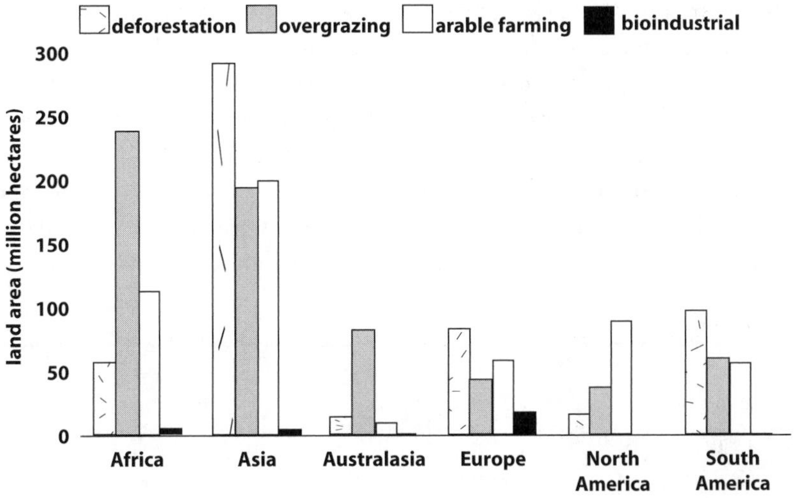

SOURCE: UN Environment Programme. 2002. *Africa Environment Outlook: Past, Present and Future Perspectives.* http://www.unep.org/aeo/253.htm. Accessed July 22, 2003.

rivers, increasing flooding risks in catchment basins, and reducing the productivity of farmland. A related problem is soil nutrient depletion, which occurs when farmers push for short-term production gains at the expense of long-term viability. National soil fertility action plans have been created in several sub-Saharan countries, but their efficacy will depend on access to soil enhancement measures (such as fertilizers, manures, compost, and bund/terrace/windbreak creation), the extent to which proposed policy reforms are implemented, and support for organic farming systems (Reardon and Shaikh 1995; UN Environment Programme 2002b).

Some countries, including Ethiopia, Kenya, Malawi, Senegal, and South Africa, have already taken steps to implement more environmentally sensitive farming and pastoral practices, and they have been rewarded with quantifiable reductions in rate of soil loss and improvements in soil fertility and productivity (Nana-Sinkam 1995). The soil fertility "action plans" utilized by these nations blend policy reforms with investment in new agricultural technologies for the dual purposes of resource conservation and economic betterment (UN Environment Programme 2002a).

Inequitable Land Rights

Land tenure reform has long been cited as a key to improving Africa's performance in agriculture and other economic sectors, but as the UN Environment Programme acknowledges, "the issue of land rights in Africa is a highly complex and sensitive social and political issue, closely linked with poverty and land degradation issues" (ibid.).

Prior to colonization, land tenure systems in Africa developed along topographical and cultural lines. Most of these systems were built on a foundation of communal access to land and other resources and communal sharing of

---

Success Stories in African Agriculture

Africa faces many challenges as it seeks to make its agricultural sector more productive, profitable, and environmentally sustainable. But experts point out that innovative farming and pastoral practices have been implemented all across the continent at regional, national, state, and local levels, with startling results. In Uganda, for example, levels of rural impoverishment dropped from 50 to 35 percent during the 1990s, after the nation embraced a series of new agricultural programs (International Food Policy Research Institute 2002).

Other recent triumphs provide hope for the future as well. For example, numerous African countries have reported surging cassava crops in the wake of breakthroughs in the eradication of the cassava mealybug, neutralization of the cassava green mite, and creation of new cassava strains. Yields have jumped by 50 percent on average, and some countries, including Nigeria and Ghana, have registered annual production growth rates of 4 percent or better throughout the 1980s and 1990s (Gabre-Madhin and Haggblade 2001; UN Food and Agriculture Organization 2002).

Similarly, African herders no longer live in fear of the dreaded rinderpest livestock disease. That disease has repeatedly eviscerated Africa's livestock populations; during one particularly deadly outbreak at the end of the nineteenth century, rinderpest killed an estimated 95 percent of the continent's cattle (Gabre-Madhin and Haggblade 2001). Rinderpest also ripped through wild ungulate populations. In the early 1980s, for example, an outbreak in Tanzania took a significant toll on buffalo, warthog, and giraffe populations. But the Organization of African Unity (OAU) established an agency to address the threat. This alliance—the Pan Africa

(continues)

benefits accrued by working the land, though each ethnic group came up with its own unique wrinkles to these land right systems. In these communities, land rights were typically passed along to subsequent generations via tradition rather than the book of law. However, the advent of the colonial era in Africa swept much of this aside, as Europeans saw the traditional systems as inefficient and impractical impediments to their own economic designs. Colonial governments subsequently claimed the land as property of the state and redistributed it as they saw fit (usually to wealthy white farmers) (ibid.).

Today, years after the last colonial administration took leave of the African continent, the colonial land tenure framework endures. But in recent decades,

Rinderpest Campaign (PARC), which eventually grew to include a coalition of thirty-five countries—developed a powerful vaccine that was subsequently distributed all across the continent. The vaccination campaign has been a resounding success, controlling the virus everywhere that it has been tried and enabling herders to increase their profits and invest in other animal health and improvement activities. The only countries where rinderpest remains a scourge are those that have been unable to mount vaccination campaigns because of civil strife and political instability (Gabre-Madhin and Haggblade 2001; UN Food and Agriculture Organization 2002).

Finally, many agriculture improvement programs introduced at the community or local level have been tremendously successful, perhaps because they have not been encumbered by the bureaucratic wrangling that typifies many efforts at national and regional levels. These programs have included crop diversification and rotation initiatives, creation of windbreaks and other erosion mitigation measures, increased emphasis on manure and other organic wastes, and sustainable irrigation schemes (UN Environment Programme 2002).

Sources:

Gabre-Madhin, Eleni, and Steven Haggblade. 2001. *Successes in African Agriculture*. Washington, DC: IFPRI.

International Food Policy Research Institute. 1997. *The World Food Situation: Recent Developments, Emerging Trends, and Long-Term Prospects*. Washington, DC: IFPRI.

———. 2002. *Ending Hunger in Africa: Only the Small Farmer Can Do It*. Washington, DC: IFPRI.

UN Environment Programme. 2002. *Africa Environmental Outlook*. Nairobi, Kenya: UNEP.

UN Food and Agriculture Organization. 2002. *The State of Food and Agriculture 2002*. Rome: FAO.

many African governments have slowly loosened their grip on their countries' fields, rangeland, and other land resources. Land reform efforts have given many black farmers their first opportunities to till their own land, and in some rural areas land reform has encouraged a renaissance in communal farming. Indeed, countries such as Tanzania, Uganda, and Mozambique have all taken steps to legally recognize community lands as legitimate (Mosely 2003). Many experts believe that these changes in land rights could be a major stepping stone toward progress in a host of policy goals, from increased investment in farming technology to improved environmental conservation. "Improved security of tenure can greatly improve land management practices, and rural development programmes should focus on greater inputs to farming, freer trade and higher value addition," claimed the UN Environment Programme. "This will ensure that: greater income is earned from production; greater food security is awarded to the household producer; and expansion of agriculture into marginal areas is controlled" (UN Environment Programme 2002a).

Increased Investment in Agriculture Programs

Investment in agricultural programs, including those focused on research, management, and production improvements, has not kept pace with need. At the beginning of the twenty-first century, African countries on average spent only 0.85 percent of their agricultural GDP on research; by contrast, industrialized countries diverted 2.6 percent of their agricultural GDP to research, which drives improvements in production and efficiency. For example, research on crop genetics has already brought improvements in drought tolerance, utilization of plant nutrients, food nutrient content, and pest and disease resistance (International Food Policy Research Institute 2002a).

Given the poor financial status of most African states—and the many programs competing for limited dollars—upgrades in agriculture investment will require a higher level of commitment from international donors, including Western governments. It has been estimated that development aid to Africa from industrialized nations has declined by one-quarter or more since the mid-1980s, when the United States, the Soviet Union, and their respective allies maneuvered for advantage in Africa. Since that time, donors ranging from the World Bank to the United States have reduced funding for agricultural programs as a percentage of their total obligations. In the case of the United States, annual spending on agricultural projects in developing countries has declined from $1.2 billion in the mid-1980s to about $400 million (Raghavan 2002).

One particularly important focus of new investment should be women farmers. Women supply more than 70 percent of the agricultural labor in sub-

Saharan Africa, in addition to being primary providers of basic family care. "Yet, agricultural researchers, extension workers, and credit providers have long neglected women's needs. When women obtain the same levels of education, experience, and farm inputs as men, they produce significantly higher yields in a range of farming systems. Designing gender-sensitive agricultural products is a win-win strategy for reducing hunger in Africa" (International Food Policy Research Institute 2002a).

African nations will also have to decide whether to pursue genetically modified strains of crops and livestock in their quest for greater food security. Certainly, the emergence of GM agriculture in other parts of the world has elicited strong reactions, both positive and negative. Advocates assert that research and investment into GM (also known as transgenic) foods could dramatically improve Africa's capacity to feed its people. They contend that the introduction of genetically modified breeds, feeds, and vaccines could make livestock herds healthier and more productive, and they note that the introduction of pest- and drought-resistant strains of maize, rice, and other crops could increase food stores while also relieving pressure on oversubscribed freshwater sources and farmland. "In some parts of Africa, draft animals cannot be used at all due to infectious or parasitic diseases. Here is where modern agribiotechnology carries special promise for the tropics: It makes possible the engineering of plants and animals (or the creation of animal vaccines) for very specific resistances to pathogens and pests" (Paarlberg 2000).

Advocates of transgenic farming believe that productivity gains associated with the introduction of GM technology would also have a transformative effect on state economies up and down the African continent. "Only when gains in agricultural productivity in poor countries bring lower staple food prices can purchases of nonfood goods finally increase, and only then can the non-agricultural part of the economy finally start to grow" (ibid.).

Proponents also say that the introduction of GM technology would have significant environmental benefits. Most notably, its capacity to help farmers produce more food on land already in use would reduce incentives to convert species-rich natural areas into fields and pastures (McHughen 2000; Pinstrup-Anderson and Schioler 2001). Given these potential benefits, analysts like the Nuffield Council on Bioethics have concluded that "the probable costs of the (mostly remote) environmental risks from GM crops to developing countries, even with no controls, do not approach the probable gains of GM crops concentrated on the local and labor-intensive production of food staples" (Nuffield Council on Bioethics 1999).

Some scientists and environmental groups, however, have expressed profound concerns about GM technology and its potential impact on public health and the environment, and they assert that Africa should pursue other avenues in increasing agricultural productivity. They urge African policy-makers to follow the lead of the European Community, which has adopted strong measures to keep GM foods out of the marketplace because of environmental and health concerns.

Many concerns about the safety of transgenic crops center on the possible interactions of genetically modified strains with the natural environment. Unplanned breeding between GM and non-GM species has been cited as a concern, since such unions could hypothetically produce insect populations resistant to the toxins in pesticides or herbicide-resistant "superweeds." Detractors also have raised concerns that consumption of GM foods could produce accumulations of toxins in human tissue (McHughen 2000; Pinstrup-Anderson and Schioler 2001).

Other people oppose the use of GM crops on economic grounds. They fear that poor African farmers will be unable to afford GM seeds, which are currently being developed and commercialized almost entirely by private firms in the United States and other economically advanced nations (Pinstrup-Anderson and Schioler 2001). Concerns have also been raised that once genetically engineered seeds are in Africa's agricultural system, cross-pollination could produce other crops containing traces of the genetic modifications, giving transnational companies a legal pathway to claim ownership of the affected crops. Finally, it has been widely noted that regulatory institutions in African countries do not currently have the capacity to evaluate and monitor the full range of possible impacts of this powerful new technology. In fact, the institutions that do exist "are prone to oversights in monitoring and implementation and outright corruption by more powerful political or private sector (including corporate transnational) actors. Keenly aware of their own internal regulatory deficits, some governments in the developing world are opting to keep GM seeds out of their farming systems entirely" (Paarlberg 2000).

Indeed, even during the 2002 famine, Malawi, Mozambique, and Zimbabwe all expressed deep concern over accepting GM food in aid packages, and the Zambian government refused these packages altogether, even though 3 million of its citizens were at risk of starvation. On the other hand, South Africa, which ranks as the economic powerhouse of the southern cone of the continent, has been strongly supportive of agricultural biotechnology. In fact, South Africa is the only country in the fourteen-nation Southern Africa Development Community (SADC) that has officially sanctioned the production of transgenic crops. Given South Africa's economic influence in the region—and its enthusiastic pursuit of agribiotechnology—other SADC countries may eventually fall in step and drop some of their objections to GM crops.

Balancing Food Production and Environmental Protection

If long-term agricultural productivity gains are to be realized in sub-Saharan Africa, the region's policy-makers, scientists, and farmers will have to address troubling trends in the stewardship of fields, rangelands, and adjacent natural areas. In many areas of Africa, pollution of land and water from agricultural activity is not nearly the problem that it is in Asia, North America, or Europe. Practitioners of subsistence farming in Africa are unable to afford the pesticides, fertilizers, and other chemical inputs that are used in other parts of the world—often so excessively that they wreak fundamental damage to river and coastal ecosystems.

In fact, some analysts believe that greater use of fertilizer and other inputs could increase food production in some areas of Africa without damaging the environment. "The land frontier is closing, making intensification—growing more on the same surface area—a critical agricultural and environmental goal," concluded one study. "Cropping intensification need not be the enemy of the environment. . . . In 'capital-led intensification,' farmers crop more intensively but offset harmful effects on soil fertility by enhancing the soil with fertilizer, manure, or compost and protecting it with bunds, terraces, and windbreaks" (Reardon and Shaikh 1995). Supporters say that instituting sustainable intensification practices on land currently cultivated would also reduce pressure on poor farmers to clear disappearing natural areas that are centers of biodiversity (ibid.).

At this point, however, sub-Saharan African states have not yet implemented environmentally sustainable models of farming. Indeed, some locales in Africa are struggling mightily with chemical contamination from commercial farming operations. In the Zambezi Basin, for example, rivers and streams have become increasingly contaminated with poisonous residues from insecticides, herbicides, and livestock feedlots (Southern Africa Development Community et al. 2000).

Sources:

Reardon, Thomas, and Asif Shaikh. 1995. "Links between Environment and Agriculture in Africa." Policy Brief No. 2. Natural Resources Policy Consultative Group for Africa (September). Available at http://www.wri.org/wri/pcg (accessed February 2003).

Southern Africa Development Community, World Conservation Union-IUCN, Southern African Research and Documentation Centre, and Zambezi River Authority. 2000. *State of the Environment in the Zambezi Basin 2000*. Harare, Zimbabwe: SADC, IUCN, SARCD, ZRA.

Tevera, D., and S. Moyo, eds. 2000. *Environmental Security in Southern Africa*. Harare, Zimbabwe: SAPES Trust.

Improving Rural Services
and Infrastructure

Rural farming communities across Africa have historically been the recipients of only modest investments in infrastructure (hospitals, schools, roads, electric service) and social services. Inevitably, these shortcomings in health and education services diminish agricultural productivity, exacerbate vulnerability to infectious diseases, and create a socioeconomic environment in which a great deal of energy is required simply to meet life's basic necessities. Limited transportation options are a particularly significant operational hurdle for farmers, as it makes it difficult for them to procure fertilizers and other inputs at affordable prices or to deliver goods to the marketplace in a timely and efficient fashion. In 2001–2002, for example, some Ethiopian farmers realized excellent harvests using high-yielding seeds and fertilizers. But the country's terrible road system made it impossible for them to get their crops to distant cities and villages. "With a glut of grain concentrated in small areas, the prices plunged by as much as 80 percent, even as other regions suffered from food shortages. As a result of low prices, farmers were unable to recoup production costs and repay loans for fertilizers. If farmers can't make money from their crops, they can't buy expensive inputs, and as a result, production is dropping dramatically" (International Food Policy Research Institute 2002b).

Agricultural performance—and quality of life—would be dramatically enhanced in these rural communities if African governments made greater investments in sustainable forms of development, from sensibly designed road systems to the introduction of renewable energy programs (including wind and solar power) to meet heating, cooking, and electricity needs. Implementation of a basic social "safety net," encompassing crop insurance, food-for-work programs, and other measures, would also be a welcome development for poor African farmers, who remain enormously vulnerable to events beyond their control, from international fluctuations in commodity prices to weather events.

Improved Institutional Performance

African states need to address widespread inefficiencies and patterns of corruption that endure not only in their agriculture and land use agencies but also in other branches of government. In addition, the nations of Africa must better integrate their agricultural and land management goals with other economic and social development goals. "Complementary policies in other sectors are needed to enhance the benefits of [agricultural] reforms and alleviate the negative effects. A stable macroeconomic environment, progress in taming corruption, and stronger legal infrastructure are prerequisites for stimulating domestic and international investment, including that in the agricultural sector. Similarly, programs to provide a credible safety net for households adversely affected by the reforms are justifiable on their own terms as well as for the political sustainability of the reforms. . . . Rural people in Africa have little chance of improving their livelihoods without well-functioning markets" (Kherallah et al. 2002).

Fortunately, many African states appear to be approaching the developmental challenges that confront them with vigor and determination. Indeed, serious proposals to address social, economic, and environmental problems in agriculture and other sectors are proliferating at the local, national, regional, and national levels, and many of these programs and policies reflect a heightened appreciation that sustainability must be a cornerstone of any future development plans. "This shift toward greater ownership of the development agenda opens the door for more countries to benefit from greater economic integration and to capture spillover benefits from the exchange of technology and information. For example, the New Partnership for Africa's Development (NEPAD) is a promising initiative among African leaders that concretely reflects the continent-wide commitment to ownership of future development priorities. And African countries are negotiating regional trading arrangements that offer new possibilities for exploiting regional dynamics, such as the Southern Africa Development Corporation (SADC) free trade area, launched in September 2000."

Such initiatives also reflect energetic efforts to build sustainable agricultural programs at the local level. "Despite restrictive public policies and grim national trends, many individuals and communities throughout Africa have taken charge of their own development and succeeded," commented one observer. "Disappointing continent-wide statistics mask gains in thousands of villages where millions of rural Africans are taking part in sustainable development activities. . . . Given the opportunity, authority, and capacity, [African farmers] have proven time and time again that they can and will adapt to new circumstances and modify their socioeconomic practices to meet both immediate needs and long-term aspirations in environmentally sound ways. Their efforts show that government inaction, state-sponsored interference, changing circumstances, and modern pressures do not always stop communities intent on managing their resources productively. If anything, the documented failure of most state-run systems to advance the lives of their citizens materially has meant that those who actually depend upon the soils, water, forests, and biodiversity for their livelihood have been compelled to devise their own systems of sustainable resource management" (Veit et al. 1995).

Sources:

Abrams, Len. 2002. "Drought and Famine in Southern Africa: A Review of the Crisis." August 11. *The Water Page,* Available at http://www.thewaterpage.com/drought_crisis_2002.htm (accessed February 2003).

Ascher, William. 2000. "Understanding Why Governments in Developing Countries Waste Natural Resources." *Environment* 42 (March).

Benneh, George. 1997. *Toward Sustainable Smallholder Agriculture in Sub-Saharan Africa.* Washington, DC: IFPRI.

Council of Arab Ministers Responsible for the Environment, UN Environment Programme, and Arab Centre for Studies of Arid Zones and Drylands. 1996. *State of Desertification in the Arab Region and the Ways and Means to Deal with It.* Damascus: ACSAD.

Gabre-Madhin, Eleni, and Steven Haggblade. 2001. *Successes in African Agriculture.* Washington, DC: IFPRI.

Gleick, Peter H. 2000. *The World's Water, 2000–2001.* Washington, DC: Island Press.

Hunter, Susan S. 2000. *Reshaping Societies: HIV/AIDS and Social Change.* Glen Falls, NY: Hunter Run.

Intergovernmental Panel on Climate Change. 2001. *Climate Change 2001: Mitigation, Impacts, Adaptation, and Vulnerability: Summaries for Policymakers.* Geneva: IPCC.

International Food Policy Research Institute. 1997. *The World Food Situation: Recent Developments, Emerging Trends, and Long-Term Prospects.* Washington, DC: IFPRI.

———. 2002a. *Ending Hunger in Africa: Only the Small Farmer Can Do It.* Washington, DC: IFPRI.

———. 2002b. *Ending the Cycle of Famine in Ethiopia.* Washington, DC: IFPRI.

Kherallah, Mylene, Christopher Delgado, Eleni Gabre-Madhin, Nicholas Minot, and Michael Johnson. 2002. *Reforming Agricultural Markets in Africa.* Washington, DC: IFPRI.

McHughen, Alan. 2000. *Pandora's Picnic Basket: The Potential and Hazards of Genetically Modified Foods.* New York: Oxford University Press.

Mosely, P. 2003. *A Painful Ascent: The Green Revolution in Africa.* London: Routledge.

Mougeot, L. J. A. *Urban Agriculture Research in Africa: Enhancing Project Impacts.* Cities Feeding People Report Series, no. 29. Ontario: IDRC.

Nana-Sinkam, S. C. 1995. *Land and Environmental Degradation and Desertification in Africa.* Rome: UN Food and Agriculture Organization.

Nuffield Council on Bioethics. 1999. *Genetically Modified Crops: The Ethical and Social Issues.* London: Nuffield Council on Bioethics.

Obasi, Godwin O. P. 2002. "Embracing Sustainability Science: The Challenges for Africa." *Environment* (May).

Paarlberg, Roger. 2000. "Promise or Peril? Genetically Modified Crops in Developing Countries." *Environment* (January/February).

Palestinian National Authority. 2000. *State of the Environment Palestine.* Gaza: Ministry of Environmental Affairs.

Pinstrup-Anderson, Per, and Ebbe Schioler. 2001. *Seeds of Contention: World Hunger and the Global Controversy over GM (Genetically Modified) Crops.* Baltimore: Johns Hopkins University Press/International Food Policy Research Institute.

Population Reference Bureau. 2002. "2002 World Population Data Sheet." Washington, DC: PRB.

Postel, Sandra. 1997. *Last Oasis: Facing Water Scarcity.* New York: Norton.

Raghavan, Sudarsan. 2002. "Starvation Stalks Africa." *Detroit Free Press* (December 2).

Reardon, Thomas, and Asif Shaikh. 1995. "Links between Environment and Agriculture in Africa." Policy Brief No. 2, Natural Resources Policy Consultative Group for Africa (September). Available at http://www.wri.org/wri/pcg (accessed February 2003).

Reich, P. F., S. T. Numbem, R. A. Almaraza, and H. Eswaran. 2001. "Land Resource Stresses and Desertification in Africa." In E. M. Bridges et al., eds., *Responses to Land Degradation.* New Delhi: Oxford University Press.

Robinson, Simon. 2002. "Scarred: War, Bad Government, and AIDS Are Feeding a Deadly Drought across Southern Africa." *Time International* (August 5).

Salloum, Habeeb. 2001. "The Flowering of Agriculture and Forestry in the United Arab Emirates." Al-Hewar Center for Arab Culture and Dialogue. Available at http://www.alhewar.com/habeeb_salloum_uae_flowering_agriculture.htm (accessed January 2003).

Southern Africa Development Community, World Conservation Union-IUCN, Southern African Research and Documentation Centre, and Zambezi River Authority. 2000. *State of the Environment in the Zambezi Basin 2000.* Harare, Zimbabwe: SADC, IUCN, SARCD, ZRA.

Tevera, D., and S. Moyo, eds. 2000. *Environmental Security in Southern Africa.* Harare, Zimbabwe: SAPES Trust.

UN Children's Fund-UNICEF. 2001. *State of the World's Children 2001.* New York: UNICEF.

UN Development Programme. 2002. *Arab Human Development Report 2002.* Available at http://www.undp.org/rbas/ahdr (accessed February 2003).

UN Economic and Social Commission for Western Asia. 1997. *Regional Report: Implementation of Agenda 21.* UN Department of Economic and Social Affairs. Available at http://www.un.org/esa/earthsummit/ecwa-cp.htm (Accessed January 2003).

UN Environment Programme. 2002a. *Africa Environmental Outlook.* Nairobi, Kenya: UNEP.

———. 2002b. *Global Environment Outlook–3 (GEO-3).* London: Earthscan.

UN Food and Agriculture Organization. 2001. *FAOSTAT 2001.* Rome: FAO.

———. 2002. *The State of Food and Agriculture 2002.* Rome: FAO.

UN World Food Program. 2003. "Africa Hunger Alert." January. Available at http://www.wfp.org (accessed March 2003).

Veit, Peter G., Adolfo Mascarenhas, and Okyeame Ampadu-Agyei. 1995. *Lessons from the Ground Up: African Development That Works.* Washington, DC: World Resources Institute.

World Bank. 2000. *World Development Indicators 2000.* Washington, DC: World Bank.

———. 2001. *African Development Indicators 2001.* Washington, DC: World Bank.

———. 2002. *World Development Indicators 2002.* Washington, DC: World Bank.

6

Freshwater

The Middle East and large sections of Africa are among the world's most arid regions. In these areas, which are characterized by sparse, seasonal rainfall and limited surface water resources (rivers and lakes), access to freshwater has long been a basic concern of human populations. Regional ecosystems, meanwhile, have evolved in accordance with the freshwater limitations that prevail across much of the continent (though not in Central Africa, which has abundant water resources). In the past half-century, the perennial pursuit of reliable freshwater has been complicated by population growth and associated development pressures in the realms of agriculture and industry, inadequate infrastructure (and financing mechanisms to improve it), poor water management practices, and political tensions. Indeed, disputes over allocations of freshwater have been cited by some analysts as a potential catalyst of armed conflict between nations, especially in the Middle East region. To date, however, the region's riparian nations—those with claims on waterways within or adjoining their territories—have been inclined to settle water disputes via diplomatic channels.

Freshwater Habitat Conservation in Africa and the Middle East

Preservation of aquatic habitats and freshwater quality is essential to any effort to preserve biodiversity and ecosystem integrity in the Middle East and Africa. Key issues in this regard include dams and other diversion projects that harness freshwater resources for human use; rising levels of pollution; and unsustainable rates of consumption.

A dam in Zimbabwe taps into the tremendous hydroelectric potential of the Zambezi River. CORBIS

Dams and Other Developments
Alter Regional Watersheds

Africa and the Middle East housed more than 1,400 large dams between them at the close of the twentieth century. According to the International Commission on Large Dams (ICOLD), Africa alone held 1,266 large dams, approximately 5 percent of the global total, while Middle Eastern states maintained another 168 large dams. In addition, both regions are believed to house a number of other dams (large and small) not included in the ICOLD inventory. Included among these totals are some of the world's most staggering architectural marvels, including Egypt's High Aswan Dam, which has the largest catchment area of any dam on the planet, at 2.2 million square kilometers; and the Kariba Dam on the Zambezi River between Zambia and Zimbabwe, which has the largest reservoir capacity—180 billion cubic meters of water—in the world (World Commission on Dams 2000). In addition, both regions contain thousands of medium-size and small dams that have been constructed for urban and rural water supply, livestock watering, and irrigation.

More than 60 percent of the large dams in Africa are located in South Africa and Zimbabwe. Most African dams were constructed in the latter decades of the twentieth century, in accordance with rising demand for water for industry, domestic use, and irrigation. Indeed, most dams in Africa have been constructed to facilitate irrigation (52 percent) and to supply water to municipalities (20 percent), although almost 20 percent of dams are multipurpose.

Ilisu Dam Project Sparks Controversy in Turkey

In the far northeastern corner of the Middle East region, Turkey has embarked on an ambitious water development project to ensure its long-term water security. This project includes plans to construct the Ilisu Dam, the largest hydroelectric dam in the nation's history, on the Tigris River. But the proposed dam has elicited heated objections from some human rights and environmental organizations.

The U.S.$1.6–$2 billion Ilisu Dam Hydroelectric Power Project is the centerpiece of a much larger scheme to build a vast network of power-generating facilities in the region. This network, known as the Southeastern Anatolia Project (known by the Turkish acronym GAP), is a massive $32 billion infrastructure development initiative that was initiated in 1978 and, if completed as planned, will place twenty-two dams and nineteen power plants on the Tigris and Euphrates rivers and various tributaries. If this work-in-progress is completed, some forecasts estimate that it will generate 27 billion kilowatt-hours of electricity, providing Turkey with hydroelectric power equal to its entire power output in 1983. This boost in power capacity will undoubtedly help Turkey meet its rising levels of power consumption; in 2001 alone it registered an 8 percent increase in consumption. Flood control and irrigation benefits will also be evident. According to some projections, wheat production could rise by 100 percent, vegetable production could jump by 40 percent, production of cotton—a crop that requires large amounts of water for cultivation—could increase by 300 percent with water drawn from GAP reservoirs. But these dams and their reservoirs will also flood vast swaths of wildlife habitat, displace numerous communities, and disrupt downstream wetlands and other habitat. Moreover, flooding will predominantly affect Turkey's Kurdish population, which has been subject to a range of state-imposed restrictions designed to snuff out rebellions seeking to establish an independent Kurdish state (Ilisu Engineering Group 2001; Ilisu Dam Campaign 2001).

As currently planned, the Ilisu Dam project alone will create a freshwater reservoir of more than 300 square kilometers (containing 10.4 billion cubic meters of water) along a mountainous stretch of the Tigris River. When the dam comes online—in 2007 or 2008, if construction schedule goals are met—it will have a capacity of 1,200 megawatts and will generate an estimated 3,800 gigawatts of power annually. The Turkish government and its supporters claim that this output will dramatically improve the living standards of communities across nine southeast Anatolia provinces.

As a consortium of European and American engineering companies begin work on the Ilisu rockfill dam, a coalition of nongovernmental organizations (NGOs) has emerged to denounce it for a host of reasons.

(continues)

Environmental groups are dismissive of the assurances of Turkish authorities, who say that the proposed project will cause only negligible ecological damage. They argue that Turkey has yet to release any meaningful assessment of the cumulative environment impact of the Ilisu Dam—or other dams in the GAP Project—because the ecological repercussions will be far from benign. These critics contend that in addition to submerging valuable wilderness areas, the dam will trap inadequately treated wastewater from upstream cities, jeopardizing regional water quality and creating a breeding ground for disease.

The Ilisu Dam Project has also been harshly criticized on humanitarian grounds. It has been estimated that as many as 183 villages and towns, inhabited mostly by minority ethnic Kurds, will be directly affected by the impoundment, with 82 settlements completely flooded and another 101 losing some portion of their land (such as agricultural fields). Estimates of the number of people who will be placed in Turkey's resettlement plan range as high as 78,000 (Ilisu Dam Campaign 2001). Moreover, detractors claim that Turkish promises of fair compensation for displacement have not been borne out at Anatolia Project dams that have thus far been completed. Opposition to the Ilisu Dam has also been strong from historians and archaeologists who lament that the reservoir will flood valuable archaeological sites dating back more than 10,000 years, including all of the town of Hasankeyf, which features a treasure trove of architecture from ancient eras. As a result, the World Archaeological Congress has lobbied the United Kingdom and other countries to withdraw financial support for the project.

Finally, many observers contend that the dam will exacerbate growing tensions with Turkey's downstream neighbors over water use. The proposed site of the Ilisu Dam is only 65 kilometers upstream from the Syrian and Iraqi borders. Turkey offers assurances that those nations will see no harm from the operations at Ilisu or other GAP facilities, but experts believe that both Syria and Iraq could lose significant volumes of much-needed water from the Tigris and Euphrates as a result of the GAP dams. Given the severe water shortages that would result in both nations, analysts see the Ilisu as a potential flashpoint for water conflict in the near future.

Sources:

Ilisu Dam Campaign; Kurdish Human Rights Project; The Corner House; World Economy, Ecology and Development; Eye on SACE Campaign; and Pacific Environment Research Center. 2001. *The Ilisu Dam, the World Commission on Dams and Export Credit Reform: The Final Report of a Fact-Finding Mission to the Ilisu Dam Region, 9–16 October 2000.* United Kingdom: Kurdish Human Rights Project.

Ilisu Engineering Group. 2001. *Ilisu Dam and HEPP Environmental Impact Assessment Report: Executive Summary.* IEG.

World Commission on Dams. 2000. *Dams and Development: A New Framework for Decisionmaking.* London: Earthscan.

Approximately 6 percent of the large dams in Africa were built primarily for electricity generation, but hydroelectric power generates more than 80 percent of the total power in eighteen African countries and more than 50 percent in twenty-five countries (World Commission on Dams 2001; UN Environment Programme 2002). The necessity of nourishing food crops has also driven dam-building in the Middle East; 56 percent of the large dams in that region were commissioned for irrigation alone or in conjunction with other purposes (World Commission on Dams 2000).

The benefits that have accrued from the creation of dams and reservoirs are significant, ranging from inexpensive—and cleaner, when compared with consumption of fossil fuels—sources of energy, enhanced agricultural productivity, and stimulation of regional economies. But dams have also been blamed for large-scale displacement of people; altered patterns of erosion and flooding; submersion of important wildlife habitat; strangulation of downstream fisheries; disruption of migratory routes for fish; and alterations to riverine ecosystems with consequent loss of habitats and species. As a result, assessing and managing the environmental impacts of dam-reservoir networks has become a higher priority in Africa, the Middle East, and other areas of the world.

Of course, other water management projects have also had an enormous influence on the character of freshwater resources. Canal systems, wetland conversions to agricultural or industrial use, and irrigation networks all have been championed for improving the living standards of nearby communities, and African economic development schemes increasingly include dam-building and wetland drainage components. But scrutiny of these initiatives has also intensified, driven by shifting perceptions about the economic impact of wetland loss and alteration on fishery-dependent communities, the quality of life enjoyed by human communities, and the long-term ramifications of such projects on natural ecosystems. "Water management projects, wherever they are located, . . . have an impact on the overall ecology of the river. . . . When the Israelis drained the Huleh Marshes in the 1950s, it was regarded at the time as a great feat of human ingenuity in converting a marshy 'wasteland' into productive agricultural land. When the Iraqis undertook the same operation in the 1990s [in the marshes of the Shatt el-Arab] current conventional wisdom accuses them of ecological devastation" (Albert 1998).

Addressing Water Pollution Problems

Freshwater quality is in decline in virtually all regions of Africa and the Middle East. The decay in water quality is attributable to myriad factors, including: heavy discharges of agrochemicals and untreated or partially treated sewage, which has caused eutrophication in lakes, reservoirs, and estuaries;

contamination of groundwater with oil, nitrates, phosphates, sewage, and salts; and loss of aquatic habitats and biodiversity. This degradation of freshwater habitats has a host of direct consequences for human communities, ranging from reductions in fish catches to increased exposure to a wide range of waterborne diseases (UN Environment Programme 2002). Less quantifiable—but still significant—is the impact that water pollutants have on the aesthetic appearance of riverside communities. In Algeria, for instance, many industries established factories directly on waterways during the 1970s and 1980s and proceeded to dump pollutants into the water at an appalling rate. "By the early 1990s . . . the level of water pollution had become disastrous. The discharging of industrial and domestic waste into rivers has reached such a scale that it is often impossible to enter any urban area without having to suffer obnoxious smells and offensive views of black, narrow waterways filled with rubbish and pollutants" (Sutton and Zaimeche 1996).

Rising Levels of Consumption

Unsustainable freshwater consumption rates prevail across most of Africa and the Middle East, especially in urban areas. In rural areas of Africa, where families expend considerable time and effort retrieving water, judicious use of the resource is encouraged. In urban areas, however, residents who need merely turn a spigot to receive water are much less conservative in using the resource. This dynamic, which is seen all around the world, exists in large part because current water consumption arrangements provide no financial incentive for conservation. Indeed, water is practically cost-free in many Middle Eastern countries, where agriculture is predicated almost entirely on irrigation systems. This remarkable state of affairs also endures in many parts of Africa, where major revisions to water policies and pricing also must be made, especially in regard to industrial and agricultural sectors that account for most freshwater consumption. Revisions are especially urgent in the semiarid countries of the Middle East and Northern and Southern Africa because climate change and continued population growth are expected to further reduce water supplies and increase water demands, respectively.

Some water conservation strategies are being pursued with vigor. "With farmers suffering crop failures in one out of every three years in many African countries, new approaches are desperately needed. Attention is turning now to the potential of smaller-scale projects—micro dams, shallow wells, low-cost pumps, moisture-conserving land techniques, and a wide variety of 'rainwater harvesting' methods. . . . Many of these efforts have proved more cost-effective and less disruptive to local communities than the large schemes that dominated development efforts during the past few decades" (Postel 1997).

Figure 6.1 Water Supply Coverage in Africa in 2000

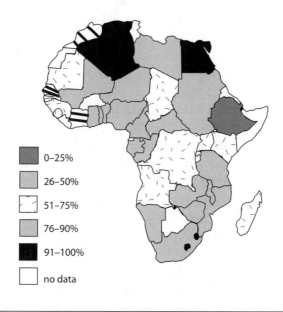

■	0–25%
▨	26–50%
▨	51–75%
▨	76–90%
■	91–100%
□	no data

SOURCE: World Health Organization. 2000. http://www.unep.org/aeo/149.htm. Accessed July 22, 2003.

Freshwater Supply and Usage
Middle East

The Middle East, also sometimes known as West Asia or the Near East, has been described as "the most concentrated region of water scarcity in the world" (ibid.). Collectively, the Middle East region—defined here as Afghanistan, Egypt, Iran, Iraq, Israel, Jordan, Kuwait, Lebanon, Oman, Saudi Arabia, Syria, Turkey, the United Arab Emirates, and Yemen—holds approximately 487 cubic kilometers of average annual internal renewable water resources. These collective holdings are less than 18 percent of the internal renewable water resources held by Canada alone (UN Development Programme 2000). Indeed, some of the region's most heralded waterways would constitute mere creeks if placed in other, water-rich environs. "The fabled Jordan River would fit comfortably in two lanes of any U.S. highway, one family sedan deep, even at peak flow," noted one observer. "The springs that feed the Jordan would barely qualify as a leak from the California Aqueduct" (Schwarzbach 1995).

The physical geography and climatic trends of the Middle East leave only Turkey, Iran, and Lebanon with generous quantities of rainfall. All other nations grapple with freshwater vulnerability on an annual basis because of rainfall that is seasonal, irregular, and localized. Some countries contain territories that are prone to drought or water shortages, while others are heavily dependent on "exogenous" waterways—those that originate outside their boundaries—that are not wholly under their control (Drake 2000). For example, Iraq and Syria are heavily dependent on the Tigris and Euphrates rivers—which originate in Turkey—for freshwater. Indeed, approximately 35 percent of the Middle East's annual renewable water resources is provided by exogenous rivers (Rogers and Lydon 1994). The strain on surface water resources has been further exacerbated by pollution. Discharges of chemicals from agricultural and industrial operations have compromised water quality in some areas and affected aquatic habitats and associated flora and fauna.

Given the paucity of surface water available across much of the region, countries such as Kuwait depend almost exclusively on groundwater aquifers for agriculture and other needs. But these resources are at risk as well, either from unsustainable rates of withdrawal or contamination stemming from human activity. For example, seepage of oil spilled during the Persian Gulf War has compromised some Kuwaiti reservoirs, and overpumping of groundwater in developed areas along the coast has caused saltwater seepage into some valuable aquifers. "The thin margin of supply that makes water resources so precious also exacerbates problems of contamination," observed one analyst. "Israeli authorities have closed a number of wells in the Coastal Aquifer due to intolerably high levels of nitrates; contamination from heavy metals and solvents is just beginning to cause concern. . . . The Gaza Strip is already facing a full-blown contamination crisis. Due to decades of overpumping from the Coastal Aquifer, the salty waters of the Mediterranean have infiltrated its southern tip. As its salinity rises, the water becomes increasingly undrinkable" (Schwarzbach 1995).

Historically, the people of the Middle East have utilized clever and farsighted practices in order to maximize their limited freshwater resources. Water conservation has been a basic tenet of communities for generations, especially in rural areas. But the task of stretching finite freshwater resources in sustainable ways has been made much more difficult in recent decades by inexorable pressure from growing populations that in turn drive the expansion of industrial, residential, and agricultural development—all of which currently require considerable volumes of water in order to be successful. In 1750, for example, the Middle Eastern population numbered around 20 million; by 1950 the populace had tripled to 60 million, and by the close of the twentieth century the population had exceeded 300 million (Population Reference Bureau 2001).

Moreover, Middle Eastern population growth has been particularly strong in urban areas, where per capita water consumption is ten to twelve times higher than in rural communities (Drake 2000). Few countries have been able to maintain sustainable models of freshwater consumption under this on-slaught. In the mid-1990s, for example, Israel was drawing on its under-ground aquifers at about 15 percent above the rate of replenishment, while Jordan was consuming about 20 percent more water than it was receiving an-nually (De Villiers 2000; Hillel 1994). These unsustainable rates of freshwater usage could undoubtedly be reduced somewhat if households, farmers, and businesses were required to pay more for this resource, but thus far, few Middle East nations have made many meaningful steps in that direction. "More water can be made available through reducing waste in irrigation, transport, distribution, and municipal and industrial uses. Far more can be done to encourage the more frugal use of water, through education and media campaigns, but especially by more realistic pricing for water use. Such changes will be hard to implement, however, and will run into enormous opposition, since water is such a politically sensitive issue; traditionally, water has been re-garded as a free resource available to all who need it" (Drake 2000).

Inevitably, concerns about water availability have added to tensions in a corner of the world that already has more than its share of ethnic, religious, and geopolitical strains. Indeed, Syrian attempts to divert the waters of the Banias River added to tensions between Syria and Israel that culminated in the 1967 Six-Day War, during which time Israel gained control over land that housed not only a large Palestinian population but also two essential water re-sources: the Golan Heights, which form the watershed of the Jordan River, and the Mountain Aquifer under the West Bank. In subsequent years, the im-portance of these lands to regional freshwater supplies has been an underap-preciated issue in peace negotiations between Israel and its Arab neighbors, all of whom see freshwater availability as a national security issue. Indeed, Israeli, Palestinian, and Jordanian leaders are fully cognizant that the freshwater re-sources they share will be insufficient to meet just the basic household and urban needs of their peoples—let alone agricultural and industrial needs—by the mid-twenty-first century, if current consumption and socioeconomic trends do not change. "This pressing reality complicates an already complex dialogue that is freighted with distrust and animosity" (Schwarzbach 1995).

Elsewhere, other riparian nations are enmeshed in contentious negotia-tions over water allocation and management issues, with each country ma-neuvering for maximum economic, legal, and national security advantage. Use of the waters of the Tigris-Euphrates, for example, has been a serious source of contention between Turkey, Syria, and Iraq. Geography has given Turkey a commanding hand in this struggle, for it owns large portions of the drainage basins of the two rivers. More than 70 percent of the combined Tigris-Euphrates water volume originates in Turkish territory, including at least 88 percent of the flow of the Euphrates, and Turkey has the capacity to store a water volume equivalent to 1.38 times the average annual flow of the river at the Syrian border. Turkey's strategically advantageous position upstream of Syria and Iraq effectively gives it the power to turn the Tigris-Euphrates spigot on and off at will, a state of affairs that is of considerable concern to the downstream riparians (Albert 1998).

In recent decades, Turkey has taken decisive steps to maximize its geographic advantage. In 1978 it initiated the Southeast Anatolia Development Project (commonly known as GAP, its Turkish acronym), an ambitious plan to construct twenty-two large dams and nineteen power plants on the Euphrates for the purpose of increasing irrigation and electricity generation and generating economic activity in an economically depressed region of the country. With the erection of each new dam and power plant, Turkey has repeated its assertion that the GAP Project will benefit Syria and Iraq by reducing damage from floods and blunting seasonal variations in water availability (Drake 2000). Turkish authorities have also offered soothing promises that its downstream neighbors will continue to receive the water they need. But Syria—which has a rapidly growing population that depends on the Euphrates for more than half of its water needs—and Iraq are unmoved by these assurances. They note that if the GAP Project is completed as planned, it will reduce the flow of the Euphrates significantly—by 30 to 50 percent by the mid-twentieth century, according to some estimates (Rogers and Lydon 1994). Furthermore, the water they do receive will be of lower quality, for it will carry salts, fertilizers, pesticides, and other agrochemicals washed into the river from irrigated fields.

Despite the growing tensions between Turkey and downstream riparians over the Tigris and Euphrates, however, other waterways have been managed with a laudable degree of transboundary cooperation. Some countries have shown a willingness to make meaningful concessions in regional water treaties in the interest of maintaining political stability and ecological integrity. Moreover, regional governments have embarked on cooperative efforts to develop advanced wastewater reclamation methods, more efficient irrigation schemes, and other means of expanding the region's water supply, 60 percent of which continues to be consumed by agriculture (Schwarzbach 1995). Finally, individual countries are pursuing desalinization and other water conservation technologies with vigor. Saudi Arabia, for example, has the greatest desalination capacity in the world, at more than 5 million cubic meters per day, and the United Arab Emirates and Kuwait rank third and fourth in the world in capacity, behind the United States (Gleick 2000). "The danger of war over water hangs over the heads of the Middle East countries, yet there is also the possibility of cooperation and harnessing new technologies and capital that would prevent such wars. Solving the water issue is one of the essential prerequisites to achieving a meaningful and lasting peace in the Middle East" (Gideon Fishelson, quoted in Wolf 1995).

Africa Overview

Africa contains approximately 9 percent of the world's total freshwater resources, but these resources are concentrated in the central and western regions of the continent, where rainfall ranges from moderate to abundant. Africa's northern and southern reaches and the Horn of Africa, on the other hand, receive comparatively little precipitation. Indeed, these regions are home to some of the planet's largest desert regions, including the vast Sahara (the world's biggest desert), which spans the midsection of North Africa; the Horn Desert in the east; and the Kalahari and Namib deserts of Southern Africa. All told, approximately two-thirds of Africa is classified as desert or arid.

The Democratic Republic of Congo-DRC (formerly Zaire), a sprawling country of verdant rain forest located in Central Africa, easily qualifies as the continent's greatest repository of freshwater. In fact, the DRC alone contains nearly one-quarter of all of Africa's annual internal renewable water resources. By contrast, Mauritania is Africa's driest country, with approximately 0.01 percent of Africa's total (UN Development Programme 2000; Gleick 2000).

Africa contains half a dozen watersheds of global significance. The northeast quadrant of the continent is home to the Nile River, the longest river in the world at approximately 6,670 kilometers (4,145 miles). Other major river basins include the Congo River in central Africa, the Niger River in western Africa, the Zambezi River in southeastern Africa, and southern Africa's Orange River. The waters of these major river systems are supplemented by underground aquifers and lake systems. Freshwater lakes in Africa have a total volume of 30,567 cubic kilometers and cover a surface area of almost 166,000 square kilometers (UN Environment Programme 2002). Most of the continent's largest lakes lie just south of the equator in the valleys of the highlands of east-central Africa. Lake Victoria—the largest lake in Africa and the third largest in the world—is located squarely on the equator, where the borders of Kenya, Tanzania, and Uganda meet. In addition, wetlands cover about 1 percent of Africa's total surface area. Wetlands of particular note include the Zaire swamps; the Sudd in the upper Nile River; the Lake Victoria basin; the Chad

Table 6.1 Major Watersheds of Africa

Major Watersheds	Modeled Watershed Area {a} (km²)	Countries within the Watershed (number)	Average Population Density (per km²)	Percent of Watershed that is: Crop-land	Forest	Grass-land	Built-up Area {b}	Irrigated Area	Arid Area	Wet-lands	Degree of River Fragmen-tation {c}
Congo	3,730,881	9	15	7.2	44.0	45.4	0.2	0.0	0.0	9.0	Medium
Cuanza	149,688	1	24	2.8	16.2	79.6	0.3	0.0	5.8	2.1	Medium
Cunene	109,832	2	10	2.6	3.3	90.9	0.1	0.1	15.8	2.9	X
Jubba	497,626	3	12	6.6	2.7	87.9	0.2	0.1	71.5	3.5	X
Lake Chad {f}	2,497,738	8	12	3.1	0.2	45.2	0.2	0.0	82.8	8.2	Low
Lake Turkana	209,096	4	60	20.8	11.9	50.2	0.1	0.3	33.0	5.9	X
Limpopo	421,123	4	32	26.3	0.7	67.7	4.5	0.9	47.3	2.8	High
Mangoky	58,851	1	18	4.5	3.3	90.8	0.1	2.3	39.1	0.2	Low
Mania	56,118	1	25	2.5	5.7	89.8	0.2	2.6	0.1	0.9	Low
Niger	2,261,741	10	32	4.4	0.9	68.6	0.5	0.1	65.4	4.1	High
Nile	3,254,853	10	44	10.7	2.0	53.0	1.0	1.4	67.4	6.1	High
Ogooué	223,946	4	2	0.8	75.1	21.7	0.5	0.0	0.0	6.2	X
Okavango {g}	721,258	4	2	5.5	1.7	91.1	0.2	0.0	75.8	4.1	X
Orange	941,351	4	11	6.0	0.2	85.0	2.2	0.5	77.0	0.8	High
Oued Draa	114,544	3	10	0.3	0.2	12.0	0.5	3.2	95.3	0.2	X
Rufiji	204,780	1	21	19.7	2.1	77.4	0.2	0.1	0.0	7.8	Low
Senegal	419,575	4	10	4.8	0.1	68.2	0.1	0.0	82.0	3.6	High
Shaballe	336,604	2	30	7.1	1.2	87.9	0.1	0.5	80.5	1.8	X
Volta	407,093	6	42	10.4	0.7	85.6	0.5	0.1	59.9	4.6	High
Zambezi	1,332,412	8	18	19.9	4.0	72.0	0.7	0.1	8.8	7.6	High

NOTES: Percentages in this table do not add up to 100 because different sources were used to estimate land cover and land use within watersheds, land cover types overlap, and not all land cover types were accounted for. "0" is either zero or less than one-half the unit measure. a. Watershed area was digitally derived from elevation data using a geographic information system; thus, area may differ from other published sources. b. Based on stable nighttime lights data. These figures overestimate the actual area lit. c. Indicates the level of modification of a river due to dams, reservoirs, interbasin transfers, and irrigation consumption. d. Countries that have <1 percent area in the watershed are excluded. e. Watershed includes intermittent tributaries in northern Chad, Niger, and Algeria. f. Watershed includes intermittent tributaries in Botswana (northern Kalahari Desert). g. Basin excludes the tidal area of the St. Lawrence River.

SOURCE: Reprinted with permission from the World Resources Institute. http://www.wri.org/wr-00-01/pdf/fw3n_2000.pdf

basin; the Okavango Delta; the Bangweulu swamps; the Kafue flats in Zambia; and the flood plains and deltas of the Niger and Zambezi rivers. As with Africa's freshwater lakes and rivers, these marshlands are essential ecosystems that sustain a large percentage of the continent's animals and plants.

At the close of the twentieth century, less than half of the available land in sub-Saharan Africa had been converted for agricultural use, and only 2.8 percent of cultivated land was irrigated (Veit 1995). Yet the agriculture sector still accounted for 63 percent of all freshwater withdrawals in Africa in 1995, and numerous countries, including Burkina Faso, Chad, Egypt, Ethiopia, Gambia, Guinea, Libya, Madagascar, Malawi, Mali, Mauritania, Morocco, Mozambique, Niger, Rwanda, Senegal, Sierra Leone, Somalia, Sudan, and Swaziland, directed more than 80 percent of their freshwater use to the agriculture sector (Shiklomanov 1999; Gleick 2000).

The continent's heavy reliance on freshwater for irrigation is expected to grow even more severe in the coming years, as population growth drives demand for food. But demand for freshwater from domestic and industrial sectors is also expected to increase, in part because of increased consumption in areas where living standards are improving. In addition, climate change associated with human activity is widely expected to advance desertification in already arid regions of the continent. Alarmed by the prospect of serious water shortages on the horizon, African authorities and international analysts have pressed for the institution of water conservation measures (such as the removal of water subsidies and the repair of leaky distribution networks), strengthening and enforcement of pollution laws, and increased management of freshwater resources on a basinwide basis. But while these goals, collectively known as Integrated Water Resource Management (IWRM), have been adopted in principle in many African countries, "implementation of IWRM has, to date, been impeded by: capacity constraints, lack of financial resources, institutional fragmentation, poor availability of information, and lack of commitment by various partners" (UN Environment Programme 2002).

North Africa

North Africa receives approximately 7 percent of the continent's total precipitation, and it houses less than 3 percent of Africa's total internal renewable water resources. In addition, distribution of precipitation is extremely uneven, with Sudan receiving about 75 percent of the region's rainfall. Yet North Africa accounts for 46 percent of the continent's total withdrawals. This disparity is a reflection both of the region's arid character and of its economic capacity to develop and maintain sophisticated water distribution schemes. The water that has been harnessed by the states of this region is consumed by a

Garbage strewn along the shoreline of Baharia Oasis, Egypt. COREL

wide variety of users, including household, industry, and tourism sectors. But as with other parts of Africa, most water is used for irrigation. In Egypt, for example, agriculture accounts for 86 percent of total water use, far outpacing consumption in the industrial (8 percent) and domestic (6 percent) sectors. Indeed, Egypt relies on irrigation to nourish virtually every square foot of land that it uses for farming (Gleick 2000).

The most visible and important body of freshwater in the region is the Nile. But considering its epic length, the Nile provides only modest water to riparian nations. In fact, its volume of discharge is dwarfed by those of the planet's other major rivers, such as the Congo, the Amazon, the Mekong, and the Mississippi. Still, its very existence in this arid climate makes it an invaluable resource—and one that has been the subject of considerable international wrangling. The Nile basin is shared by ten countries, but historically, Egypt and Sudan have accounted for virtually all withdrawals. Indeed, according to the terms of the 1959 Treaty for the Full Utilization of the Nile, the two countries divided the river's annual flow between them, with 66 percent allocated to Egypt, 22 percent to Sudan, and the remainder set aside for surface evaporation and seepage at Egypt's immense Aswan High Dam (AHD) reservoir. "No other riparians were party to this accord, nor was any share in the annual flow prudentially set aside for them. The treaty did anticipate future demands from the excluded riparians, but merely stipulated that Egypt and Sudan should deal with those demands jointly. Predictably, the upstream riparians, many at the time still under British control, were not pleased. However, Cold War rivalries, their own political instability combined with bouts of civil war, and their feeble economies did not allow upstream riparians to press any realistic claims to Nile water arising in or traversing their territories" (Albert 1998).

In the 1990s, however, economic and political conditions improved in places like Uganda, Eritrea, Ethiopia, and Tanzania, and these upper riparians may soon decide to challenge the Egypt-Sudan arrangement in order to expand their own irrigation efforts, which are inextricably intertwined with efforts to feed growing populations (Waterbury 1997). Increased capture of river water by Ethiopia and other upper riparians would constitute a serious threat to food and economic security for Sudan and, especially, Egypt, which is growing at a rate of about 1 million people annually and relies on land irrigated by Nile water for virtually all of its agricultural production. Egypt has expressed particular concern about the designs of Ethiopia, where about 85 percent of the river's flow into Egypt originates. In 1993, Egypt and Ethiopia signed an agreement covering the Nile's water resources. However, Ethiopia has repeatedly expressed interest in expanding its hydroelectric and irrigation capacity through new projects that, if completed, would materially reduce the volume of water received by Egypt (Drake 2000). Analysts also believe that Sudan, located immediately upstream from Egypt, might well insist on a greater share of Nile water if it ever emerges from the quicksand of civil war and associated socioeconomic impoverishment. "What will happen when Ethiopia and Sudan begin demanding more of the Nile's water? Will Egypt accept the Helsinki and International Law Commission rules that irrefutably entitle Ethiopia and Sudan to a larger portion of Nile water? Will Egypt try to change those rules to give greater weight to the principle of prior use? Or will it be tempted to use its position as the most powerful nation in the Nile basin to assure its present allocation, even if this means the use of military force and international conflict?" (ibid.).

In an effort to avoid bleak scenarios of military action over water, the ten riparian nations of the Nile launched a joint Nile Basin Initiative (NBI) in 1999. Chief objectives of the NBI include sustainable resource development, economic security and integration, and transboundary cooperation in management of the watershed's resources. In May 2003, NBI countries gathered to issue a joint "Dar es Salaam" declaration endorsing a cooperative approach to hydroelectric power generation and other utilization of shared water resources.

Across much of the rest of the region, surface water resources are extremely modest and the Sahara Desert holds sway. As a consequence, reliance on groundwater aquifers (most notably the Nubian Sandstone Aquifer and the Continental Intercalaire Aquifer) is considerable. Libya, for instance, gets 95 percent of its freshwater from groundwater resources. But communities along the Mediterranean coastline, where the vast majority of this region's populace lives, are draining water tables at a pace that is outstripping natural rates of re-plenishment and causing saltwater intrusions in some coastal aquifers. In Morocco, for instance, major hotels are a notorious drain on regional aquifers. But cities such as Agadir are so dependent on tourism that they are loath to place limits on water use. As a result, the city is exploiting water from underground tables located in the surrounding Souss Plain, which is already under immense pressure from fruit and vegetable growers. The aquifer is now being drained at a phenomenal rate—about 60 million meters annually in ex-cess of the aquifer's recharge rate. Yet Agadir continues to expand its battery of hotels and construct other water-intensive developments to attract tourists. Given these trends, it is unlikely that Agadir and other like-minded communi-ties will be able to avoid acute water shortages in the future (Berriane 1996).

The chief water quality-problems facing North Africa include salinization from poor irrigation methods, pollution from industrial and municipal sources, saltwater intrusion in coastal aquifers, and eutrophication (a condi-tion in which excess organic nutrients from fertilizers and waste deplete a body of water's supply of oxygen). In Egypt, for instance, domestic wastewater from hundreds of towns and villages and dozens of factories fouls the waters of the Nile, as do vast quantities of chemical-laced water washed from heavily treated farm fields. In Tunisia, meanwhile, the deposition of agricultural pes-ticides, household waste, and sewage into water resources is a major problem, even though the government has invested in sewage infrastructure improve-ments and worked to relieve water demand through expanded reuse of treated wastewater. In the early 1990s, for example, Tunisia's households were releas-ing approximately 100 million cubic meters of untreated wastewater annually (Henchi 1996).

State agencies and international organizations are pursuing a range of pro-grams to improve the quality and conserve the quantity of water in North Africa. These measures include new water sanitation schemes, investments in water monitoring programs, new environmental protection laws, and in-creased integration of water quality and aquatic habitat concerns into other policy areas (UN Environment Programme 2002).

West Africa

The sixteen nations of West Africa exist under a wide range of climatic condi-tions. Those in the southernmost sector contain large swaths of outright rain forest, albeit with lesser levels of endemism and biodiversity than the forests of Central Africa. These regions generally receive moderate to heavy rainfall and have adequate freshwater resources for their needs. Countries such as Mauritania, Niger, and Chad, on the other hand, contain large areas of desert within their borders. Communities in these nations of the southern Sahara frequently grapple with drought, and desertification is a looming ecologic concern. In fact, acute water shortages have been predicted for Benin, Burkina Faso, Ghana, Mauritania, Niger, and Nigeria by 2025. Of all the West African countries, Liberia has the greatest amount of internal renewable water resources, at more than 63,000 cubic meters per capita per year. Mauritania, by contrast, has only 150 cubic meters per capita per year. Availability of groundwater also varies considerably (ibid.).

In addition to increased desertification linked to global climate change, major factors implicated in West Africa's dwindling water supply include population growth, rapid expansion of commercial agriculture operations, and industrial development. In fact, agriculture accounted for 70 percent of all withdrawals in the region at the close of the twentieth century, and some analysts believe that this percentage could grow even greater, since some fertile farming areas have not yet been exploited. Finally, degradation of rivers, lakes, and groundwater resources is a serious problem in much of the region, further aggravating freshwater shortages in some communities. This pollution, which takes the form of industrial effluents, untreated human and livestock waste, and, to a lesser degree, fertilizer and pesticide runoff from fields, has taken a particularly heavy toll on the region's coastal areas. Native flora and fauna in places such as the Korle Lagoon in Ghana have been devastated by a host of pollutants that have wreaked havoc on their aquatic habitat, and offshore environments have been compromised by pollutants carried out to sea (ibid.).

At this juncture, obstacles to more effective management and dispersal of West Africa's freshwater resources has been hamstrung by meager funding, political instability, deteriorating infrastructure, and legal and social structures that fail to provide meaningful incentives for the protection and conservation of rivers, aquifers, and other water sources. But gains are being realized in some areas. Ghana, Niger, Mauritania, and other countries have installed new water treatment and sanitation programs, and many West African states have banded together with other countries on the continent to explore improvements in water distribution, pricing, and protection. In addition, West African states have added dozens of wetlands to the RAMSAR List of Wetlands of International Importance, which are managed with an emphasis on preserving ecological integrity, biological diversity, and hydrological functions. In 2002, for example, the government of Guinea announced the designation of a

4.5 million hectare swath of wetlands—including the headwaters of the Niger River—under the RAMSAR banner. This area thus became the second-largest area of protected wetlands in Africa, after Botswana's Okavango Delta.

East Africa

As a whole, East Africa features a moderate supply of freshwater, with total average renewable freshwater resources amounting to 187 cubic kilometers per year (ibid.). But freshwater availability, quality, and use is highly variable in the region. Some countries have been able to dam large waterways to provide for much of their electricity generation and irrigation needs, while simultaneously providing safe drinking water to their citizens. In Djibouti, for example, water supply and sanitation coverage is 100 percent in both urban and rural areas (World Health Organization and UN Children's Fund 2000). Other states, however, are so parched that they are racked by drought-induced famine on a depressingly regular basis. The arid countries situated in the Horn of Africa—Eritrea, Ethiopia, and Somalia—have been the primary victims of drought and famine in the past half-century, while Uganda receives the highest levels of renewable freshwater. Groundwater is only a minor component of total freshwater supplies across much of the region.

The river systems of East Africa have been radically transformed by human intervention in the form of hydroelectric dams, reservoirs, and diversion networks for irrigation (Ward 2002). In Kenya, for example, nearly 80 percent of its electricity is generated by hydroelectric dams. But not all states have been able to effectively harness available freshwater sources for their people. In Ethiopia, for example, fewer than one out of four people have access to clean water, even though the nation has numerous lakes and rivers within its borders (World Health Organization and UN Children's Fund 2000).

Access to freshwater is also complicated by the region's highly seasonal rainfall. For example, Ethiopia typically receives 75 percent of its rain in a three-to-four-month period. In some countries, this annual cycle of flooding and drought casts a pall of uncertainty over the availability of water for electricity generation, irrigation, drinking, and industrial consumption. Indeed, during periods of water scarcity, demand now frequently exceeds available supply. This state of affairs is likely to intensify in future years, as population growth spurs increased demand for water for drinking and irrigation.

Land development pressures stemming from rising populations also are placing some wetlands, rivers, and lakes at risk. Overgrazing by large herds of livestock has degraded and depleted water resources at some lakes and rivers, both by increasing the susceptibility of riverbanks to erosion—which in turn creates sedimentation that degrades aquatic habitat—and by contaminating waterways with waste. Poor agricultural practices also have dumped high volumes of agrochemicals in waterways, exposing humans, animals, and plants to harmful toxins and causing eutrophication of many bodies of water. In Lake Victoria, for example, increased concentrations of nitrogen and phosphorus washed down from surrounding tea and coffee plantations have not only contributed to the alteration of species composition but also made lake conditions ideal for the explosive growth of water hyacinth, which is bedeviling many rivers and lakes in East Africa. This outbreak has altered aquatic ecosystems, reduced the quality of water, and clogged commercially vital navigation channels (UN Environment Programme 2002).

But farming is not the only culprit in the region's declining water quality. Domestic sewage and stormwater runoff are major freshwater pollutants in myriad cities, as are industrial facilities—many of which are located directly on rivers or estuaries—that discharge effluents laced with heavy metals and other contaminants into waterways after only minimal treatment (most discharges are not treated at all). This pollution has been blamed for the devastation of aquatic ecosystems and marked declines in various animal and plant species that depend on clean water for their survival. In 2001, for example, the famous flamingos that adorn the shores of lakes Nakuru and Bogoria in Kenya died in droves; subsequent research traced their demise to deadly exposure to heavy metals deposited in the lakes (ibid.). In addition, although major dam-reservoir networks provide communities with needed electricity and water for agricultural operations, their impact on river ecosystems has come under increased scrutiny in recent years.

In response to growing concern about the quality and availability of freshwater and the integrity of freshwater habitats, several East African states have taken tangible steps to manage their resources in a more sustainable manner. Kenya, Ethiopia, and Uganda, for example, have all formulated policies to better protect and manage their existing freshwater resources. But funding for such measures remains problematic across most of the region, and major shortfalls in physical and institutional infrastructure, enforcement of environmental laws, and scientific monitoring of freshwater resources continue to persist.

Central Africa

Most of the nations of Central Africa are blessed with high levels of rainfall and numerous rivers and lakes. Indeed, this is the most water-rich region of the African continent. In 1998 the region's annual withdrawal of freshwater for all purposes was estimated to be less than 1 percent of the total available, and precipitation is so abundant and reliable that agriculture accounts for only one-third of all withdrawals, a far lower percentage than in other areas of Africa, where arid conditions prevail (UN Development Programme 2000). In addition, Central Africa features a lower population density than most other areas of the continent. Consequently, many of the region's waterways remain relatively unspoiled by pollution or other forms of human alteration such as

Desalination: The Wave of the Future?

Desalination—a technological process whereby salt is removed from saltwater so as to make it suitable for drinking, crop cultivation, and other purposes—has emerged as a tantalizing tool in the planet's increasingly desperate quest to find enough freshwater for all its people. Since 97 percent of all water on earth is saltwater, the removal of salt from saltwater would provide the planet's human inhabitants with a virtually limitless supply of water for drinking and crop cultivation. But despite its technical feasibility and its growing use in some arid regions of the world—most notably the Middle East—desalination makes only a modest contribution to the planet's overall water supply. In the late 1990s, for example, the total amount of desalinated water produced in an entire year was equivalent to the amount the world used in fourteen hours (Gleick 2000).

Commercially viable desalination plants remove salt from saltwater (or brackish groundwater) either through distillation—heating water and condensing the steam—or through reverse osmosis—filtering the water through a membrane. Both of these methods effectively produce water that is perfectly fine for drinking or irrigation. But both modes of production, and especially distillation, require a great deal of energy. This reality has raised concerns about greenhouse gas emissions generated by desalination activity. In addition, it has made desalination among the world's most expensive water supply options. Large-scale desalination operations are thus beyond the financial reach of many countries, including some developing nations already facing severe freshwater shortages.

In the oil-rich regions of the Middle East and North Africa, however, desalination is an integral element of many water supplies. Indeed, more than half of the world's total desalination capacity is in the Middle East/Arabian Gulf/North Africa regions, with approximately one-quarter of global capacity concentrated in Saudi Arabia alone. Other countries with considerable desalination capacity include Kuwait and the United Arab

(continues)

dams. The banks of the mighty Congo River, for example, are largely devoid of development for most of the waterway's length, enabling it to avoid the level of degradation that has befallen so many other major waterways in Africa and other parts of the world.

Nonetheless, competition for water is accelerating between nations and among various constituencies (agriculture, industry, municipal), and pollution is intensifying in rivers and especially coastal estuaries, where pollutants

Emirates, which rank third and fourth in the world, behind the Saudis and the United States (ibid.).

Saudi Arabia and other Middle East countries with ample energy supplies are now using desalinated water for almost all of their domestic consumption. Opinions vary, however, about the technology's future in other parts of the world. According to some analysts, desalination stands as "the only realistic hope" for dealing with freshwater shortages. "A desalination project for 10,000 people costs the equivalent of one military tank; for 100,000 people, the price is roughly that of a jet fighter. Investing in desalination of brackish water or sea water . . . is cheaper than attempts to settle disputes over available water sources, most of them already overused" (Nachmani 1997). But other experts, while granting that it is a boon to some regions, believe that the technology's high cost precludes it from widespread use. They contend that less expensive water conservation measures such as wastewater recycling, rainwater harvesting, and the use of low-flush toilets hold greater promise. "Desalination remains a solution of

last resort. The frequency with which cities and communities are turning to it is more a sign of water scarcity and stress than it is a source of comfort. . . . Desalination will be an expensive life saver to a growing number of coastal cities and towns bumping up against supply limits, but it does not constitute the oasis needed in the global water picture. Its costs are way out of line with what farmers, the world's biggest water users, can pay. As with the megadiversion projects that engineers dream of, desalination holds out the unrealistic hope of a supply-side solution, which delays the onset of the water efficiency revolution so urgently needed" (Postel 1997).

Sources:

De Villiers, Marq. 2000. *Water: The Fate of Our Most Precious Resource.* New York: Houghton Mifflin.

Gleick, Peter H. 2000. *The World's Water, 2000–2001.* Washington, DC: Island.

Nachmani, Amikam. 1997. "Water Jitters in the Middle East." *Studies in Conflict and Terrorism* 20 (January–March).

Postel, Sandra. 1997. *Last Oasis: Facing Water Scarcity.* New York: W. W. Norton.

generated by industrial and agricultural operations and urban and rural households are accumulating at a troubling rate. Saltwater intrusion into groundwater supplies is also becoming more commonplace in coastal areas. Heavy rates of logging in regional watersheds have compromised aquatic habitats, as soil previously held in place by root systems is washed into rivers and estuaries and waters previously shaded by forest canopies undergo temperature changes.

Finally, global warming has been cited as a potential threat to vital freshwater resources. The Lake Chad basin, for instance, provides seven countries with water for drinking and cultivation of crops, and it is an important fishery. But the lake has undergone significant shrinkage in recent decades, and the Intergovernmental Panel on Climate Change (IPCC) warns that reduced rainfall and other climatic changes associated with global warming could increase regional desertification and further reduce the volume of water carried by rivers entering the lake. These developments have the potential to wreak ecologically devastating reductions in the lake's size (Intergovernmental Panel on Climate Change 2001).

Southern Africa

The nations of southern Africa are at a pivotal point in their management of freshwater resources, as accelerating demand is crashing headlong into immutable limitations in supply. Certainly, the African continent's southern reaches are not bereft of freshwater resources. However, the type, quantity, and quality of freshwater vary wildly across the region, seasonal differences in water availability are considerable, and disparities in access to freshwater are great. In Botswana, for example, 95 percent of the population has ready access to water, while in war-torn Angola, the percentage is only 38 percent (World Health Organization and UN Children's Fund 2000).

Among southern Africa's major freshwater resources are Lakes Tanganyika and Malawi, the world's second- and third-deepest freshwater lakes; the Zambezi and Orange river basins; and significant groundwater aquifers. Indeed, groundwater is the main source of water for about 60 percent of both urban and rural people in southern Africa, and in some places, like Botswana, groundwater accounts for 80 percent or more of the water consumed by humans and animals (Chenje and Johnson 1994).

But groundwater sources are under assault on a number of fronts. The most pressing threat is an unsustainable rate of withdrawal that pushes regional water tables down every year. Significant groundwater end-users range from urban residents to mining operations, but agricultural operations dependent on groundwater for irrigating crops rank as the single biggest consumer. In South Africa, for instance, irrigation accounts for approximately three out of four gallons of water consumed, even though only 1 percent of the land inside its borders is irrigated. This irrigated land accounts for 30 percent of the value of the state's total agricultural production, a welcome rate of productivity given the steady population growth evident throughout the southern end of the continent. But according to some estimates, as much as 50 percent of the water used by South African farmers is lost through leaking pipes and evaporation. This inefficiency also can trigger serious environmental problems ranging from soil poisoning through salinization to outright waterlogging of land. In addition, irrigation water is often used to produce low-value crops that would not be economically viable without substantial state subsidies (Chenje and Johnson 1994).

Southern Africa's rivers, many of which are transboundary waterways with marked seasonal variations in flow, are also in jeopardy. Heavy levels of withdrawal for irrigation have exacerbated seasonal shortages in some communities, particularly among downstream riparian nations, and water pollution is a major issue in many locales. In urban areas, poor sewage treatment and disposal facilities are the rule rather than the exception, producing waterways that are suffused with all manner of contaminants, from human waste to oil. Each year, for example, Zambia's Kafue River receives nearly 93,000 tons of waste generated by an assortment of chemical, textile, and fertilizer plants along its shores (Chenje 2000). In the case of Mozambique, more than 120 factories in and around Maputo lacked waste treatment plants in the mid-1990s; as a result, these facilities discharged toxic wastes, poisons, and waste matter directly into the city's neighborhoods (Chenje and Johnson 1996).

Not surprisingly, waterways that have been degraded in this manner have proven a potent incubator for waterborne diseases. In rural areas, meanwhile, fertilizer and pesticide use is often quite low by international standards. But in some locales, rainfall washes agrochemicals from treated fields into rivers and streams, where they cause eutrophication and poison aquatic flora and fauna. Mining activities are another significant contributor to declining water quality and degradation of freshwater habitats. The rivers in Greater Soweto, for example, have been heavily contaminated by drainage water from mining operations (Greater Johannesburg Metropolitan Council 1999). Exploitive logging and other poor land-management practices have also dumped high loads of sediment into rivers, where it blocks navigation channels, ruins fish spawning beds, and increases vulnerability to flooding.

Finally, the rivers of southern Africa have been irrevocably changed and reshaped by extensive networks of dams, reservoirs, and canals. According to the World Commission on Dams, this region features the highest concentration of dams and interbasin transfer schemes anywhere on the continent (World Commission on Dams 2001). This infrastructure has provided important benefits in terms of electricity generation and crop cultivation. But dams and canal networks have also fundamentally altered riverine habitat (by interrupting fish migrations, for instance) and disrupted seasonal replenishment of flood plains that serve as productive fish nurseries and a source of nutritious grassland for wildlife and livestock (Chenje and Johnson 1994). Nonetheless, pressure to approve and construct infrastructure capable of transferring water from underutilized areas to areas of high demand is so great that assessments of the environmental repercussions of such activities are mere afterthoughts. For example, construction of southern Africa's first transboundary water transfer—the Lesotho Highlands Water Project (LHWP)—began before environmental impact assessments were even published (Masundire 1993).

Analysts worry that southern Africa received a stark glimpse into its future in the first few years of the twenty-first century, when drought conditions devastated several countries in the region, most notably Algeria, Lesotho, Malawi, Mozambique, Swaziland, Zambia, and Zimbabwe. This 2001–2002 drought—compounded by economic problems, population displacement stemming from war and civil unrest, depleted food reserves, and ineffectual government responses—has raised the specter of famine for an estimated 16 million people. In recognition of the region's extreme vulnerability to freshwater shortages, countries such as Botswana, Namibia, and South Africa have mounted meaningful efforts to create policies that reflect environmental realities. South Africa, in particular, has embraced the concept of ecosystems as legitimate users of water, passing a series of progressive reforms in the realm of water policy.

Sources:

Abrams, Len. "Drought and Famine in Southern Africa: A Review of the Crisis." The Water Page, available at www.thewaterpage.com/drought_crisis_2002.htm (accessed October 28, 2002).

Albert, Jeff, et al. 1998. *Transformations of Middle Eastern Natural Environments: Legacies and Lessons.* New Haven, CT: Yale University Press.

Alden Wily, L. A., and S. Mbaya. 2001. *Land, People and Forests in Eastern and Southern Africa at the Beginning of the 21st Century.* Nairobi: World Conservation Union.

Berriane, Mohamed. 1996. "Environmental Impacts of Tourism along the Moroccan Coast." In Will D. Swearingen and Abdellatif Bencherifa, eds., *The North African Environment at Risk.* Boulder, CO: Westview Press.

Biswas, A. K. 1994. *International Waters of the Middle East—From Euphrates-Tigris to Nile.* Oxford: Oxford University Press.

Central African Regional Program for the Environment. 1996. *CARPE Workshop, Libreville, Gabon.* Washington, DC: USAID.

Chenje, Munyaradzi, ed. 2000. *State of the Environment: Zambezi Basin.* Maseru: SADC, IUCN, ZRA, SARDC.

Chenje, Munyaradzi, and Phyllis Johnson, eds. 1994. *State of the Environment in Southern Africa.* Harare, Zimbabwe: South African Research and Documentation Centre, IUCN-World Conservation Union, and Southern African Development Community.

———, eds. 1996. *Water in Southern Africa.* Harare, Zimbabwe: South African Research and Documentation Centre, IUCN-World Conservation Union, and Southern African Development Community.

DeVilliers, Marq. 2000. *Water: The Fate of Our Most Precious Resource.* New York: Houghton Mifflin.

Drake, Christine. 2000. "Water Resource Conflicts in the Middle East." *World and I* 15 (September).

Gleick, Peter H. 2000. *The World's Water, 2000–2001.* Washington, DC: Island.

Gorbachev, Mikhail, and Shimon Peres. 2000. "Water: The Key Issue of the 21st Century." *New Perspectives Quarterly* 17 (Summer).

Greater Johannesburg Metropolitan Council. 1999. *State of the Environment Johannesburg.* Johannesburg, South Africa: GJMC.

Henchi, Belgacem. 1996. "Pollution and the Deteriorating Quality of Life in Tunisia." In Will D. Swearingen and Abdellatif Bencherifa, eds., *The North African Environment at Risk.* Boulder, CO: Westview.

Hillel, Daniel. 1994. *The Rivers of Eden: The Struggle for Water and the Quest for Peace in the Middle East.* Oxford: Oxford University Press.

Intergovernmental Panel on Climate Change. 2001. *Climate Change 2001: Impacts, Adaptation and Vulnerability: Summary for Policymakers.* Geneva: IPCC.

Lee, Michael D., and Jan Teun Visscher. 1990. *Water Harvesting in Five African Countries.* Delft: International Water and Sanitation Centre.

Masundire, H. 1993. "Large Dams and Large-Scale Water Transfers." Harare, Zimbabwe: Southern African Research and Documentation Centre.

Naff, Thomas. 1997. "Information, Water, and Conflict: Exploring the Linkages in the Middle East." *Water International* 22 (March 1).

Ohlsson, Leif, ed. 1995. *Hydropolitics: Conflicts over Water as a Development Constraint.* London: Zed.

Population Reference Bureau. 2001. *2001 World Population Data Sheet.* Washington, DC: PRB.

Postel, Sandra. 1997. *Last Oasis: Facing Water Scarcity.* New York: W. W. Norton.

———. 1999. *Pillar of Sand: Can the Irrigation Miracle Last?* New York: W. W. Norton.

Revenga, Carmen, et al. 1998. *Watersheds of the World: Ecological Value and Vulnerability.* Washington, DC: World Resources Institute/Worldwatch Institute.

———. 2000. *Pilot Analysis of Global Ecosystems: Freshwater Systems.* Washington, DC: World Resources Institute.

Rogers, P., and P. Lydon. 1994. *Water in the Arab World: Perspectives and Prognoses.* Cambridge: Harvard University Press.

Schwarzbach, David A. 1995. "Promised Land." *Amicus Journal* 17 (summer).

Shiklomanov, I. A. 1999. *World Water Resources: Modern Assessment and Outlook for the 21st Century.* St. Petersburg: Federal Service of Russia for Hydrometeorology and Environment Monitoring, State Hydrological Institute.

Sutton, Keith, and Salah Zaimeche. 1996. "Desertification and Degradation of Algeria's Environmental Resources." In Will D. Swearingen and Abdellatif Bencherifa, eds., *The North African Environment at Risk.* Boulder, CO: Westview.

Uitto, Juha I., and Jutta Schneider, eds. 1997. *Freshwater Resources in Arid Lands. UNU Global Environmental Forum V.* Tokyo: United Nations University Press.

UN Development Programme, UN Environment Programme, World Bank, and World Resources Institute. 2000. *World Resources 2000–2001: People and Ecosystems, The Fraying Web of Life.* Washington, DC: World Resources Institute.

UN Environment Programme. 2002. *Africa Environment Outlook.* Hertfordshire, UK: Earthprint Limited and UNEP.

UN Food and Agriculture Organization. "FAO's Information System on Water and Agriculture: General Summary, Near East Region," Available at www.fao.org/ag/agl/aglw/aquastat/regions/neast.

———. 1997. *Water Resources of the Near East Region: A Review.* Rome: FAO, Aquistat Programme.

Veit, Peter G., Adolfo Mascarenhas, and Okyeame Ampadu-Agyei. 1995. *Lessons from the Ground Up: African Development That Works.* Washington, DC: World Resources Institute.

Ward, Diane Raines. 2002. *Water Wars: Drought, Flood, Folly, and the Politics of Thirst.* New York: Riverhead.

Waterbury, John. 1997. "Is the Status Quo in the Nile Basin Viable?" *Brown Journal of World Affairs* 4, no. 1.

Wolf, Aaron. 1995. *Hydropolitics along the Jordan River: Scarce Water and Its Impact on the Arab-Israeli Conflict.* Tokyo: UN University Press.

World Commission on Dams. 2000. "Large Dams and Their Alternatives in Africa and the Middle East: Experiences and Lessons Learned," Available at www.dams.org/kbase/consultations/afrme.

———. 2001. "Africa: Irrigation and Hydropower Have Been the Main Drivers for Dam Building. Dams and Water: Global Statistics," Available at www.dams.org.

World Health Organization and UN Children's Fund. 2000. *Global Water Supply and Sanitation Assessment 2000 Report.* Geneva: WHO.

World Water Council. 2000. *The Africa Water Vision for 2025: Equitable and Sustainable Use of Water for Socio-economic Development.* Presented at the 2d World Water Forum in The Hague, The Netherlands, March. Available at http://watercouncil.org/Vision/Documents/AfricaVision.PDF.

Oceans and
Coastal Areas

Much of Africa and the Middle East is characterized by arid climate and associated freshwater limitations. Indeed, human communities from the Atlas Mountains of northern Africa to the Cape of Good Hope at the southern tip of the continent maintain societies built upon the acknowledgement that water is a precious resource to be husbanded. Yet these predominantly parched lands are virtually surrounded by the vast waters of the Atlantic and Indian oceans and three inland seas—the Red Sea, Mediterranean Sea, and Black Sea—that are tenuously connected to those larger oceans. These marine waters are an invaluable resource to the region's human communities, and they have profoundly influenced the development of regional cultures and economies. They provide fish, shrimp, and other species to subsistence fishermen and commercial fleets alike (indeed, fish is almost the sole source of animal protein in many artisanal communities), and they serve important economic functions as tourist destinations and shipping routes. In the case of Middle Eastern countries such as Saudi Arabia and the United Arab Emirates, which are pursuing desalination technologies, these seas even serve as an important source of water for domestic, agricultural, and industrial use.

Moreover, the coastal and offshore waters of Africa and the Middle East provide habitat for a splendid array of flora and fauna, from the fish that dart through the Gulf of Guinea to the endangered dugong that roam the saltwater marshes of the East African coast. However, the integrity of many of the marine and coastal ecosystems upon which these creatures rely is under considerable pressure from a host of human activities. These activities include overfishing, particularly by commercial enterprises; alteration and degradation of mangrove forests, coral reefs, and other vital coastal habitats;

and contamination of estuaries and offshore waters with pollutants that are generated hundreds of miles inland by destructive agricultural, logging, and industrial practices and then carried downstream. Efforts to mitigate the destructive impact of overfishing, pollution, and coastal development on the environment have proliferated in Africa and the Middle East in recent years, driven in part by increased recognition of the long-term repercussions of ecosystem degradation on human social and economic systems. But implementation of meaningful marine and coastal protection measures has been uneven to date, constrained in large part by the economic and political instability that dogs many nations in this part of the world.

Fisheries Threatened by Pollution, Overfishing by Foreign Fleets

Periodic upwellings of nutrient-rich currents in the eastern Atlantic and western Indian oceans are the cornerstone of this region's fishery productivity. These currents provide sustenance for large populations of seabirds, deepwater species such as tuna, sardines, and mackerel, crustaceans such as shrimp and crab, and myriad species of coastal fish.

This marine life is essential to many coastal communities where artisanal fishing is central to local diets and economies. But in many places declining fish stocks have placed this traditional lifestyle in jeopardy. One culprit in this downturn is degradation of coastal areas that serve as important feeding, breeding, and nursery areas for fish. Another factor is the predatory practices of some "distant water" multinational fishing fleets, which utilize the latest technological innovations—and employ environmentally destructive practices like bottom trawling—to meet the soaring global demand for seafood and fish products.

Overexploitation of finite fisheries resources is hardly limited to Africa and the Middle East. Indeed, 70 percent of the world's fish stocks were classified as fully exploited or overexploited at the close of the twentieth century. But the consequences will be particularly severe for this region if it fails to address overfishing trends, for some of the fish stocks undergoing the most severe depletion are dietary staples. Per capita availability of fish in Africa has declined since the 1970s, and in countries such as Ghana and Liberia, the average diet contained less fish protein in the 1990s than it did during the 1970s. Already, artisanal fishermen desperate to fill empty nets are increasingly raiding marine national parks and other protected areas, or placing their lives at risk by venturing past depleted near-shore areas into deeper waters far from shore (UN Food and Agriculture Organization 2002; UN Environment Programme 2002a).

Despite growing concerns about dwindling fishing stocks, many African governments continue to see the sale of fishing rights to foreign fleets as an easy source of income. Bidders for these rights are not in short supply. Indeed, flags from all over the world can be found fluttering above the decks of the trawlers crawling across the eastern Atlantic and western Indian oceans. For example, the European Union, which has been grappling with dwindling fish stocks in its own waters, had bought fishing rights from fifteen African countries as of 2002. The EU has vigorously defended these agreements, contending that they contain significant provisions for local employment and pointing out that the agreements help fund programs that promote sustainable fishing. But these agreements have been criticized by conservationists for allegedly weak enforcement of catch limits and depletion of vulnerable fish stocks that are essential to poor African coastal communities. Critics also claim that the financial terms of the agreements are enormously advantageous to the European Union, given the value of targeted fish stocks on the world market.

Marine Fisheries in the Middle East and Northern Africa

The principal marine waters of Northern Africa and the Middle East—the western Atlantic Ocean, the Mediterranean and Black seas, the Red Sea, and the Arabian Sea and its Persian Gulf—contain economically important fisheries, some of which have given up higher catch volumes in recent years. For example, catches by the nations of Northern Africa, which ply the waters of the mid-Atlantic and the southern Mediterranean Sea, rose by about 30 percent from 1990 to 1997, from 845,000 tons to 1.1 million tons (UN Food and Agriculture Organization 2001).

The fisheries off the Atlantic coast of Morocco remain the region's biggest prize, although commercial fleets from Morocco and elsewhere are plying these waters in such great numbers that overexploitation of target species such as sardine and mackerel has emerged as a concern. But the region's higher catch volumes are also attributable to increased fishing activity in the Mediterranean, where harvests of many species have risen despite serious concerns about resource degradation from water pollution and coastal development. Indeed, lucrative fishing grounds exist in Tunisia's Gulf of Gabes, Libya's Gulf of Syrta, and Egypt's Nile Delta region. On the whole, though, the Mediterranean's northern waters, which receive heavy infusions of nutrients from a number of rivers, are more productive than the sea's southern reaches (UN Food and Agriculture Organization 1996b). Moreover, some nations with claims on southern Mediterranean waters have reported troubling fish catch trends. For example, in Israel, which takes the great majority of its fish from the Mediterranean, landings have shown a more or less continuous decline over the last two decades, "a sure sign of over-exploitation of the stocks" (UN Food and Agriculture Organization 1996b).

Fishing is not a major cog in the national economies of any of the countries of this region, although Morocco has made considerable investments in industrial processing and export of fish products. But fisheries do provide an important source of employment and income in some coastal pockets, and in some isolated communities people depend almost entirely on fish and fish products for their livelihood (ibid.).

These subsistence communities are at the mercy of industrial, urban, and agricultural forces that in recent decades have deposited vast amounts of toxins and effluents into coastal waters and inland waterways (which subsequently carry pollutants to the sea). This environmental degradation is placing many marine ecosystems at risk. Indeed, outbreaks of jellyfish and algae that thrive in nutrient-saturated waters have emerged as a summertime scourge along long swaths of the Mediterranean coastline. In addition, the combjelly that decimated marine ecosystems in the Black Sea has established a foothold in the waters of the northeastern Mediterranean (ibid.).

In the Persian Gulf (also known as the Arabian Gulf), all commercially valuable fish stocks are fully exploited. Catch volumes during the mid-1990s sagged from levels reported in earlier years, but it has been speculated that side effects of the 1991 Persian Gulf War—such as destruction of coastal shrimp nurseries and oil spills—might be partly responsible for the downturn. Another frequently cited culprit in declining fish stocks is feverish coastal development. For example, inland developments such as Iraq's Third River Irrigation Network and Turkey's massive Southeast Anatolia Development Project (GAP) hydroelectric project threaten to disrupt discharges of nutrient-rich freshwater upon which the gulf's marine species rely. "The discovery and exploitation of oil in the Gulf region in 1901 changed its fortune, and the coastal landscape as well," summarized one analysis. "Rapid urbanization necessitated coastal installations, development of ports, construction of power and desalination plants and other activities attendant with oil wealth. All continue to impact the Gulf environment" (Sheppard 2000).

In the northern Arabian Sea and its Gulf of Aden and Gulf of Oman arms, fish stocks are threatened by industrial fishing fleets, many of which hail from Southeast Asia, Northern Europe, and other distant ports. Enforcement of catch quotas, seasonal restrictions, licensing requirements, and other regulations is anemic, and reliance on environmentally destructive fishing techniques

A farmer gathers a net at a fish farm in Israel. MICHAEL BUSSELLE/CORBIS

is standard on many trawlers. For example, vessels roaming the Gulf of Oman and the larger Arabian Sea in search of surface-dwelling tuna make extensive use of long-lining and purse-seining equipment that is notorious for snaring and killing large numbers of seabirds, dolphins, and other "by-catch"—non-target species. Similar trawling activities are also commonplace in the Gulf of Aden, where populations of valuable species like cuttlefish and lobster have been harvested to the point of collapse (Sheppard 2000; MacAlister Elliott and Partners 1998).

Aquaculture accounts for a modest percentage of total fish production in the region. According to FAO figures, total aquaculture production within the region amounted to only 0.6 percent and 2.2 percent of world production by volume and value, respectively, in the mid-1990s. But while aquaculture operations accounted for only 5 percent of total fish production in the region as a whole, they made major contributions to individual countries. For example, aquaculture accounted for 70 percent of Israel's fish production and 50 percent of Syria's production (UN Food and Agriculture Organization 1996b). The popularity of aquaculture in these countries reflects widely held convictions that the industry can relieve pressure on wild fisheries while simultaneously supplementing the dietary needs of human communities. But the ecological impact of aquaculture operations has come under severe scrutiny. Critics contend that fragile coastal habitats have been disrupted to make way for fish pens, and that wild species of fish are being exposed to diseases and waste generated in aquaculture pens. Concerns have also been raised that cross-breeding of escaped cultured fish with wild stock might reduce the survival capacity of the latter.

Marine Fisheries in Sub-Saharan Africa

The coastal nations of sub-Saharan Africa have made extensive use of their substantial marine fisheries over the last thirty years. This trend is especially evident on the western coast of the continent, as countries bordering the Atlantic Ocean account for about 85 to 90 percent of the region's total marine domestic landings in most years (approximately 70 percent of freshwater catches are taken from the eastern side of the continent, which features numerous large and productive lakes). Marked differences in fishery health exist within this coastal area, however. The northern and southern reaches of the Atlantic coast are characterized by abundant resources and low population densities, while the fisheries of the Gulf of Guinea are shrinking under heavy population pressure (UN Food and Agriculture Organization 1996a).

At the close of the 1990s, most fleets, foreign and domestic, were concentrating their energies on three major fishing areas: the eastern central Atlantic, from Mauritania to the Democratic Republic of Congo; the southeast Atlantic, along the shores of Angola, Namibia, and South Africa; and the western Indian Ocean, including the Red Sea. The outlook for targeted stocks varies across these and other fishing grounds of sub-Saharan Africa. All told, however, most commercially valuable bottom-dwelling species are believed to be fully exploited, prompting speculation that lightly exploited open-water species may garner increased attention from fleets in the coming years (UN Food and Agriculture Organization 2002).

Regional fisheries have always played an important role in providing communities with food and employment, especially in rural areas. Indeed, approximately 8 million people (about 20 percent of the region's total agriculture workforce) are directly or indirectly involved in the sector, including an estimated 1 million artisanal fishers who cast their nets into the bays and islands of the Atlantic and Indian oceans. In the West African nations of Mauritania, Guinea-Bissau, and Senegal alone, it is estimated that half a million people depend directly on fisheries for their food and livelihoods. High levels of employment in this sector are attributable to increased investments in the fishing industry, which have enabled the industry to emerge as a significant source of foreign exchange. Indeed, countries such as Mauritania, Senegal, Madagascar, Namibia, Mali, Ghana, Seychelles, and Mozambique all received more than 5 percent of their total gross domestic product or foreign currency earnings from commercial fishing in the late 1990s. During this period, the shrimp fishery on Mozambique's Sofala Bank accounted for 40 percent of the country's total foreign exchange all by itself (UN Food and Agriculture Organization 1996a.)

Despite marked increases in investment and catch volumes, the domestic production of sub-Saharan Africa remains a modest contributor to global fishery production, accounting for less than 4 percent of world totals through the mid-1990s. It should be noted, however, that these figures do not include catch from foreign fleets, which are a significant presence in African waters. All told, distant-water fishing nations are believed to account for 25 to 35 percent of all the fish and other seafood taken out of African fishing grounds (UN Food and Agriculture Organization 1996a).

Fish and seafood pulled from the ocean still accounts for a significant percentage of the African population's total animal protein intake—about 15 to 20 percent in recent years. In some subsistence communities, this percentage is much higher. But while per capita fish consumption in sub-Saharan Africa hovers above the world average, it has declined in recent years, a product of declining fish stocks and the diversion of a growing proportion of the total marine catch to the European Union—which accounts for as much as 85 percent of total fish exports—and other foreign markets (ibid.).

The above statistics show broad trends in sub-Saharan Africa's fisheries. But in some cases, they mask considerable regional variations in outlook. In East Africa, for example, efforts to tap into potentially lucrative offshore fisheries in a sustainable manner have been stymied by war, poverty, and political instability and corruption. Mismanagement of fishery resources is so complete and widespread that it has created a dynamic wherein most coastal fish stocks are considered to be fully exploited or overexploited, yet the coastal fishery yield for the entire eastern and southeastern African coast and the island states of the Western Indian Ocean accounts for less than 1 percent of worldwide landings (UN Food and Agriculture Organization 2002).

Indeed, marine ecosystems along several sections of the East African coastline have been pulverized by widespread use of dynamite, poisons, purse-seine nets, and other destructive fishing methods, as well as indiscriminate harvesting of juvenile fish. All of East Africa's coastal nations bear some responsibility for this state of affairs. For example, regional shark populations have plummeted as a direct result of illegal shark-fin catches by fishermen from Yemen, Somalia, Djibouti, and Sudan. The grim situation has also been exacerbated by incursions on the part of foreign fleets hailing from Japan, Taiwan, France, Spain, and South Korea. These trawlers have operated with particular impunity in Somalia's virtually lawless waters and also have taken a heavy toll on fish stocks in the Exclusive Economic Zones (EEZs) of other nations. Under the UN Law of the Sea, all coastal nations have sovereign control over the waters and seafloor that lie up to 12 miles off their shores, as well as dominion over seas extending 200 miles from inhabitable land; the latter is known as its Exclusive Economic Zone (UN Environment Programme 2002a; Sherman 1998).

In West Africa, meanwhile, the implementation of national fisheries management schemes have driven steady growth in catch volume since the 1950s. Harvests of small but commercially valuable open water species such as anchovies, herring, and sardines have been an integral part of this expansion, accounting for almost half of the total catch volume in recent years (UN Environment Programme 2002a). The allure of these commercially valuable fisheries, however, has prompted foreign fleets to descend on these waters in growing numbers. Mindful of the financial straits of many West African nations, these fleets are able to negotiate advantageous fishing agreements that permit them to take large harvests in exchange for relatively modest payments. These arrangements hurt West African trawler operators and subsistence fishermen, neither of whom can match the technology and other resources brought to bear by vessels from Europe and elsewhere. Conservationists contend that this heavy pressure has resulted in drastic reductions in the populations of some commercially valuable fish stocks and degraded marine habitat to the point that populations of dolphins, sharks, turtles, and other species have been affected (World Wide Fund for Nature 2002).

In Central Africa, where the Gulf of Guinea attracts pressure from artisanal fishers, regional fishing vessels, and foreign fleets, exploitation of both deep water and coastal species is approaching—and in some cases exceeding—sustainable levels. In order to prevent disastrous declines in valuable fishery stocks and associated damage to marine ecosystems, Central African governments have been urged to negotiate more environmentally sensitive fishing agreements with foreign fleets and devote greater resources to punishing illegal fishing (such as exceeding catch limits or fishing species out of season) that threatens the livelihoods of local fishermen (UN Food and Agriculture Organization 2002).

Signs of unsustainable fishing pressure also are mounting in Southern Africa. These trends—reduction in the volume of catches, decreases in the mean sizes of fish, and the like—will be difficult to reverse. But unlike other coastal areas of sub-Saharan Africa, some nations in this sector already have significant fishery monitoring and management institutions in place. Indeed, regulations on size limits, bag limits, closed seasons, and no-fishing regulations in marine reserves and other protected areas have all been imposed, although enforcement of these measures in some areas is inadequate (ibid.). Within the region, South Africa has taken a leadership role in fishery and marine habitat conservation. "South Africa's coastal and marine resources are under considerable threat and are already severely degraded in many areas due to over-harvesting and urban/industrial development," explained the South African Department of Environmental Affairs and Tourism. "Unless management of ecosystem integrity and coastal sensitivity to development is improved immediately, these resources will be lost for good" (South African Department of Environmental Affairs and Tourism 1999). In 1998 it passed the Marine Living Resources Act, which dictates that all South African fish stocks must be used on a sustainable basis and ends harvesting of overexploited populations until they have recovered to sustainable levels. In December 2000, grim reports of dwindling fish populations convinced the government to declare a temporary suspension of all fishing—including fishing by artisanal and sport anglers as well as commercial fishers—to allow stocks to recover. In addition, the government has touted the creation of a National Coastal Management Policy that makes important provisions for integrated coastal management and sustainable use of fisheries and other resources, and it has created forty-four marine protected areas in which harvesting of fish stocks and other resources is forbidden or strictly regulated (South African Department of Environmental Affairs and Tourism 2000).

If the other nations of sub-Saharan Africa hope to reverse disquieting trends in the fortunes of the fish and other marine life that inhabit their waters, analysts believe that they will have to adopt sustainability-oriented measures similar to those propagated by South Africa, and provide the necessary funding for their implementation. In addition, the thirty-six African nations that have signed the 1982 UN Convention on the Law of the Sea (UNCLOS), which came into force in 1997, must fulfill their obligations under that treaty, including legally binding provisions for the protection and preservation of marine environments. Finally, African states have been urged by marine scientists, environmentalists, and other constituencies to negotiate more economically and environmentally beneficial agreements with foreign fleets, some of which have been criticized for unsustainable levels of exploitation and endangering the livelihoods of artisanal fishers and their families (UN Environment Programme 2002a).

Marine Resources and the Threat of Pollution

Africa and the Middle East possess a wonderful array of marine areas and coastal environments, including mangrove forests, dunes, saltwater marshes, lagoons and bays, coral reefs, and rocky shores. These varied habitats support a wealth of biodiversity, both marine and terrestrial. But in many sectors of this region, the ecological integrity of these natural systems is at risk not only from unsustainable harvests, as discussed above, but also from pollution generated by factories, cities and smaller settlements, agriculture, and other human activities. Contributing causes to these festering pollution problems include impoverishment and state corruption; political instability; inadequate awareness of pollution's negative impact on the environment, public health, and economic development; weak environmental policies and institutions; and nonexistent or deficient legal mechanisms for enforcing environmental laws. "The most effective means of controlling coastal and marine pollution and degradation [are] therefore to demonstrate to industry and the public the benefits of maintaining a healthy environment—for example, improved aesthetic qualities and leisure facilities, improved harvests, and increased revenues from tourism" (ibid.).

Pollution in the Middle East

The seas surrounding the Arabian Peninsula—the Mediterranean and Red seas and the Persian Gulf (also known as the Arabian Gulf)—are particularly susceptible to pollution because of their semienclosed status. Since they have only limited tidal currents and saltwater exchanges with larger seas, which would help them flush out pollution from inland and coastal sources, pollution in these waters tends to become concentrated over time.

This heightened vulnerability to pollution has resulted in the endangerment of many species of flora and fauna. The Red Sea, for example, contains some of the world's most northerly coral reefs, and these reefs—composed of nearly 200 distinct species of coral—are home to at least 450 reef-associated species, including many endemic ones (species found nowhere else in the world) (Wilkinson 2000). In addition, mangrove forests gird long sections of Red Sea coastline, providing shelter and nursery habitat for crab, shrimp, and migratory waterbirds, and seagrass beds provide sustenance for endangered turtles and dugongs.

Fortunately, human activity along the Red Sea is concentrated in a few major cities, which has kept long stretches of coastline in fairly natural states. But the shores of the Red Sea are still the site of a great deal of human activity, from industrial development to tourism. These activities have contaminated the Egyptian coast, sections of the Gulf of Suez, and other parts of the Red Sea with a smorgasbord of pollutants, including industrial toxins, human waste, hazardous wastes dumped from ships, oil spills and leaks, and agrochemicals. For example, the deposition of large volumes of untreated sewage by Yemen and other countries bordering the Red Sea has damaged coral reef ecosystems, degraded coastal areas, and contributed to eutrophication problems (eutrophication is a phenomenon in which excessive concentrations of nitrates and phosphates spawn feverish plant growth; in some areas of the world, this phenomenon has been so severe that oxygen is depleted and marine life is faced with suffocation or relocation). Individually, each of these pollution sources constitutes a potentially major threat to the environment. Combined, these forces, if left unaddressed, have the capacity to wreak havoc on habitat and species (Hinrichsen 1998; Sheppard 2000).

On the northeastern side of the Arabian Peninsula, meanwhile, oil pollution in the waters of the Persian Gulf is considerably higher than the world average, due to events such as the 1983 Nowruz oil spill during the Iran-Iraq War and contamination from the 1991 Gulf War. Indeed, despite the Gulf War's short duration (six weeks), it had devastating environmental consequences for the region. For example, one oil slick from the war extended more than 90 kilometers long and 15 kilometers wide, affecting coastal and marine habitats from the offshore coral islands of Saudi Arabia to the intertidal marshlands of Kuwait. In addition, the brief but violent conflict deposited hundreds of ships on the sea bottom, including some with holds full of crude oil.

To the west, water quality in the Mediterranean Sea has been eroded by the collective impact of twenty-one coastal nations with a collective population of more than 400 million, including major industrialized countries such as Egypt, Turkey, Italy, and France. In addition, much of the regional economy revolves around tourism, which dramatically heightens the volume of waste generated. This is no small environmental consideration, given the fact that many cities and towns along the Mediterranean dump their municipal sewage into the sea in untreated or partially treated form. In addition, marine pollution from industrial and agricultural sectors has long been a serious problem in many littoral nations. For instance, high levels of nutrients contained in agricultural runoff have been cited as a factor in rising incidences of eutrophication, and estimates of the annual discharges of oil into the sea—most of it from routine operations rather than disastrous spills—have ranged from 500,000 to more than 700,000 metric tons. These statistics indicate that the Mediterranean, which accounts for only 1 percent of the world's ocean surface, receives almost 20 percent of all oil spilled or discharged in the world's oceans (Hinrichsen 1998; European Environment Agency 1998).

Pollution in Africa

Water quality in both coastal and offshore areas is poor in many regions of Africa. Pollution enters the water from numerous sources, and the volume of waste generated by most of these sources is considerable. One of the chief culprits in the degradation of Africa's marine environment is huge discharges of untreated industrial waste and sewage from rapidly expanding coastal settlements into marine waters or surrounding rivers and wetlands. In some cities perched along the Gulf of Guinea, for example, waste from less than 2 percent of households is treated before it is dumped into the sea (UN Industrial Development Organization, undated). In South Africa, dozens of sewage outfalls located along the coast discharge approximately 800,000 cubic meters of sewage and industrial effluent into the sea every day. Most of these large pipelines discharge into deeper waters, but more than two dozen older pipelines discharge above the high water mark (South African Department of Environmental Affairs and Tourism 1999). Massive garbage dumps located in sensitive coastal areas are another growing problem, for they not only degrade surrounding coastal habitat but also spill into coastal waters as a result of wind, wave, and rain action. Plastic products account for a growing percentage of the material in these dumps, which is problematic at several levels. First, plastic is not biodegradable. Second, it poses a mortality threat to shorebirds and other animals that ingest or become entangled in it.

High rates of population growth have also encouraged heavy industrial development in coastal regions, and local and state authorities have failed to rein in environmentally destructive practices that are commonplace in these commercial zones. These breweries, textile industries, fish processors, slaughterhouses and tanneries, petroleum processing facilities, chemical and manufacturing facilities, and aluminum smelters are notorious for discharging effluents directly into rivers, lakes, lagoons, and coastal waters (Hinrichsen 1998). Conservationists are also concerned that the discovery of new oil reserves off the coast of West Africa is spurring new exploration and drilling activities that could be damaging to the marine environment.

Offshore dumping of toxic wastes and other contaminants by the shipping industry is another serious problem, especially in high-traffic areas like the Cape of Good Hope, a region prone to stormy and unpredictable weather that nonetheless sees 4,000 oceangoing transport vessels pass through on an annual basis, generating as much as $U.S.500 million every year (South African Department of Environmental Affairs and Tourism 1999). "Waters around Africa are major transportation routes for oil and there have been many serious accidents in recent years, including the break up of the *Apollo Sea* in 1994 and the *Treasure* in 2000, both off the Cape of Good Hope," confirmed the UN Environment Programme. "However, it is not just oil spills that are a problem. Oil tankers frequently empty ballast and wash engines on the high seas and residues of degraded oil are consolidated and washed onto the shores by wind currents and waves. . . . Added to this are leaks from oil drilling activities, port handling of oil and petroleum products and refineries located in coastal zones, and leaks from barges, tankers, pleasure craft and fishing boats" (UN Environment Programme 2002a). In addition, seabed mining for diamonds, titanium, and other minerals is a growing industry in some regions of Africa. These offshore activities benefit regional and national economies and spur job creation, but they have also been criticized for disturbing coastal wetlands, estuaries, and sand dune systems; diminishing local air quality; and altering the habitat of bottom-dwelling species (South African Department of Environmental Affairs and Tourism 1999).

Finally, resource exploitation in inland areas has had a visible impact on the ecological health of some estuaries and bays. For example, environmentally destructive logging practices in river basins increase sediment loads in waterways, which can disrupt passage for migratory fish and smother spawning beds, and residues from fertilizers and pesticides used in farming activities are carried by rivers to the sea, where they contribute to algal blooms, toxic red tides, and other manifestations of eutrophication. Dams, on the other hand, prevent normal sediment loads from reaching rivermouths, which deprives coastal zones of nutrients upon which marine species depend and heightens the river's scouring potential, causing higher rates of erosion in coastal zones. In Ghana, for example, the Akosombo Dam has been blamed for coastal erosion that has reached six meters per year, while dam systems in the Upper Niger, Volta, and Benue rivers have wreaked major changes on the character of the Niger Delta (UN Environment Programme 1999).

The cumulative impact of this marine pollution is so great that countless wild ecosystems—and many species of flora and fauna dependent on those ecosystems—have already been irrevocably damaged. Moreover, degradation of coastal and offshore areas has had a pronounced impact on the economic fortunes of some cities and villages. Specific problems include contamination of shellfish through red tides, deposition of heavy metals from industrial effluents, increased incidence of cholera, typhoid, and hepatitis as the result of pollution of coastal waters by sewage, and accumulations of plastics and other solid wastes in areas of tourism (UN Environment Programme 2002a).

Coastal Areas in Decline

The coastlines of many parts of Africa and the Middle East are under heavy pressure from growing human populations and associated industrial and agricultural development activities. In Northern Africa, for example, 40 to 50 percent of the population in Mediterranean countries lives in coastal areas, with

Figure 7.1 Growth of Cities and Industry Threatens Coastal Environments

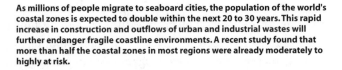

As millions of people migrate to seaboard cities, the population of the world's coastal zones is expected to double within the next 20 to 30 years. This rapid increase in construction and outflows of urban and industrial wastes will further endanger fragile coastline environments. A recent study found that more than half the coastal zones in most regions were already moderately to highly at risk.

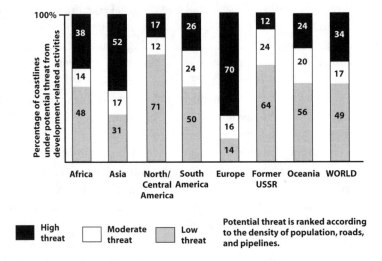

SOURCE: UN Food and Agriculture Organization. 1998. http://www.fao.org/news/factfile/FF9804-E.HTM. Accessed July 22, 2003.

population densities that approach 1,000 people per square kilometer in some urban areas (World Bank 2001). In West Africa, meanwhile, it has been estimated that one-third of the total population is clustered along a 60-kilometer-wide swath of coastal land between Senegal and Cameroon (UN Environment Programme 2002b). In some regions, heavy development pressure is further exacerbated by seasonal influxes of tourists.

High rates of coastal erosion are an unfortunate reality in many countries. Island states in the Western Indian Ocean are at particular risk from this phenomenon as a result of self-inflicted wounds. Poorly regulated mining of coral, sand, limestone, and shells has eviscerated large sections of the coastal buffer zones that previously protected these islands from storm surges and wave action over countless millennia. In Mauritius, for example, it has been estimated that laborers excavate (and transport via dugout canoes) a million tons of coral sand from nearshore areas every year (Wilkinson 2000). On the African mainland, meanwhile, construction of ports, harbors, beachfront developments, agricultural plantations, and other developments have altered significant expanses of shoreline. "Commercial developments such as hotels

Alexandria, Egypt, is one of the many coastal cities experiencing rapid population growth. PHOTODISC, INC.

and resorts have been constructed rapidly and widely along the coast of Eastern Africa, from southern Sudan to Kenya, often without adequate planning or provision of basic services such as waste disposal and sanitation," observed one report. "Many developments are concentrated on beaches and sea fronts, to benefit from the clean beaches, clean lagoon waters and healthy coral reefs, and to attract as many visitors as possible" (UN Environment Programme 2002a). Mining of sand, gravel, coral, and lime from estuaries, beaches, and shallow bays further contributes to accelerated erosion rates. Coasts have even been affected by activities taking place hundreds of miles inland, for logging, dam-building, and clearance of natural areas for farming and other purposes all influence the flow and sediment loads of rivers emptying into area seas.

The Potential Impact of Climate Change on Africa

Forecasts of impending climate change cast Africa as a victim. The continent contributes negligible amounts of greenhouse gases associated with global warming, but rising temperatures and sea levels predicted under authoritative climate change scenarios would deliver crushing blows to Africa's social, economic, and institutional foundations, as well as its environment. For example, the Intergovernmental Panel on Climate Change (IPCC) cites a one-meter rise in global sea levels by 2100 as a distinct possibility. In Alexandria, Egypt, alone—a city in which the present population of 3 million is predicted to more than double by 2030—a sea level rise of one meter could displace as many as half of its residents while simultaneously obliterating its finest beaches, its most hallowed cultural and historical monuments, and much of its commercial, industrial, and residential infrastructure. If similar scenarios unfolded in other coastal population centers, the potential consequences for human communities and flora and fauna alike are almost impossible to fathom (Intergovernmental Panel on Climate Change 1998).

Cities, villages, and plantations in low-lying coastal areas along the Gulf of Guinea, Senegal, Gambia, Egypt, island states located in the Western Indian Ocean, and coastlines along much of East Africa would be most severely affected by rising sea levels, with contamination of already oversubscribed freshwater sources and massive displacement of human populations among the most dire problems. But the ripple effect of rising sea levels would extend into inland states as well, in the form of large refugee populations and increased competition for extremely finite natural resources.

If global sea levels do rise by one meter in the coming decades, the archipelago that makes up the Seychelles would lose approximately 70 percent of its land area. On the West African coast, a region characterized by lagoons, deltas, and other low-lying areas, a one-meter jump in sea level would inundate 18,000 square kilometers of heavily populated coastline and submerge mangrove forests that are cornerstones of regional biodiversity. In North Africa, submersion of the Nile Delta, which accounts for 45 percent of Egypt's agricultural production and 60 percent of its fish production, would constitute a serious blow to ecosystems and human communities alike. And in the southeastern reaches of the continent, rising sea levels would bury the mangroves of Tanzania and Mozambique and obliterate coastal urban areas such as Cape Town and Dar es Salaam (Intergovernmental Panel on Climate Change 1998 and 2001).

It is also believed that global climate change associated with greenhouse gas emissions will warm surface water temperatures in the oceans surrounding Africa. These temperatures have been forecast to rise more modestly (about 0.6 to 0.8 degree Celsius) than waters in other parts of the world, but the impact on marine ecosystems would still be significant. In the western Indian Ocean, for example, the 1998 El Nino phenomenon boosted sea temperatures by more than 1 degree Celsius; this increase was cited as the chief factor in the bleaching deaths of between 50 and 90 percent of the corals off the Kenyan coast. Further south, warmer sea temperatures were also cited as the primary culprit in the loss of up to 90 percent of the coral reefs in some coastal areas of Tanzania, Mozambique, and South Africa, including numerous reefs that provided vital habitat for commercially valuable fisheries

(Wilkinson 2000). Scientists even speculate that rising surface-water temperatures could increase the frequency and intensity of hurricanes and other severe storm events in the region.

Restoring and Protecting
Ocean Waters and Coastal Zones

The level of marine degradation in many areas of Africa is distressing and must be addressed quickly if imperiled ecosystems are to be saved. But saving the continent's oceans and coastlines is not a hopeless cause. Some marine and coastal areas remain relatively healthy, providing conservationists with anchors on which to attach restoration efforts. For example, Kenya's mangrove forests still provide vital nursery habitat for a wide range of species, and the reefs hidden beneath the waters off the coasts of Djibouti, Eritrea, and Somalia still contain the highest levels of diversity of coral and other reef species in the entire Indian Ocean. In addition, many African and Middle Eastern states have exhibited a greater willingness to address marine and coastal protection issues in recent years.

For example, regional initiatives aimed at responding to marine and coastal pollution have proliferated over the past two decades. International agreements of note in which African and Middle Eastern nations participate include the Convention for the Prevention of Pollution from Ships (MARPOL), the Regional Convention for the Conservation of the Red Sea and Gulf of Aden Environment (Jeddah Convention), the Barcelona Convention's Mediterranean Action Plan, the Nairobi Convention for Eastern Africa, and the Abidjan Convention, an emergency response and pollution mitigation plan also known as the Convention for Cooperation in the Protection and Development of the Marine and Coastal Environment of the West and Central African Region.

Some of these regional initiatives are paying visible dividends. For example, coastal states on the Gulf of Guinea such as Benin, Cote d'Ivoire, Ghana, Nigeria, Togo, and Cameroon are all members of an ocean protection program funded by the UN Development Program and other international agencies. This effort—the Gulf of Guinea Large Marine Ecosystem Programme—has established regional effluent regulations and standards for participating countries and introduced a pilot industrial waste management program. Municipal water pollution control, sustainable development, and biodiversity conservation are other focuses of the program, which ultimately aims to improve the health of coastal waters all along the gulf (UN Environment Programme 2002a).

In the realm of coastal zone protection, several countries are placing greater emphasis on mangrove and reef rehabilitation and preservation projects, and environmental impact studies are now required prior to coastal development in Egypt, Gambia, Ghana, Kenya, Mauritius, Nigeria, South Africa, Swaziland, Tanzania, Uganda, Zambia, and Zimbabwe. Africa is also participating more fully in the international debate over global warming, and taking a more active role in seeking viable solutions to that looming issue. Transboundary cooperation in coastal zone management is also on the upswing, which has produced new programs in the areas of sustainable agriculture, soil conservation, coastal erosion mitigation, and watershed management. One pioneering example of this integrated approach to resource management can be found in southwestern Africa, where the nations of Angola, Namibia, and South Africa, all of which border on the Benguela Current Large Marine Ecosystem, have pioneered new efforts to integrate management of their respective coastal zones in environmentally sustainable ways (O'Toole 2001). Finally, some individual countries are being proactive in addressing shortcomings in their stewardship of marine and coastal resources. Egypt, for example, is not only a signatory to all regional and international agreements designed to protect seas and coastlines but is also working to establish sustainable models of ecotourism and coastal development on its holdings along both the Red Sea and the Mediterranean.

Despite these and other heartening examples of heightened environmental consciousness, however, major new investments of time, money, and resources must be made across Africa and the Middle East if the region hopes to preserve its seas and coastal areas for future generations. Existing pollution mitigation policies and environmental conservation targets that have been allowed to wither on the vine must be fully funded, and major new monitoring and conservation initiatives have to be implemented. "The environment is under siege," concluded the UN Environment Programme in remarks that covered the entire world, but are particularly applicable to the seas and coasts of the Middle East and Africa. "Unless both short- and long-term changes are instigated, sustainable development will remain a chimera—possibly only in the haze on a distant horizon" (UN Environment Programme 2002b). These keys to sustainable development of coastal and marine resources are unlikely to take place, however, if states do not make marked progress in alleviating poverty, improving governance, and eliminating debt burdens that currently encourage developing nations in Africa and elsewhere to overexploit their environmental resources.

Sources:

Bennun, L. A., R. A. Aman, and S. A. Crafter, eds. 1992. *Conservation of Biodiversity in Africa: Local Initiatives and Institutional Roles.* Nairobi, Kenya: National Museums of Kenya.

Chenje, Munyaradzi, and Phyllis Johnson, eds. 1996. *Water in Southern Africa.* Harare, Zimbabwe: South African Research and Documentation Centre, IUCN-World Conservation Union, and Southern African Development Community.

Commission for Sustainable Development. 1999. *Country Reports to the 7th Session of the Commission for Sustainable Development, New York.* Available at http://www.un.org/esa/agenda21/natlinfo (accessed December 1, 2002).

European Environment Agency. 1998. *Europe's Environment: The Second Assessment.* London: Elsevier.

Goni, Raquel, Nicholas V. Polunin, and Serge Planes. 2000. "The Mediterranean: Marine Protected Areas and the Recovery of a Large Marine Ecosystem." *Environmental Conservation* 27 (June).

Groombridge, Brian, and Martin D. Jenkins, in association with the UN Environment Programme and World Conservation Monitoring Centre. 2002. *World Atlas of Biodiversity.* Berkeley: University of California Press.

Hinrichsen, Don. 1998. *Coastal Waters of the World: Trends, Threats and Strategies.* Washington, DC: Island.

Intergovernmental Panel on Climate Change. 1998. *The Regional Impacts of Climate Change: An Assessment of Vulnerability.* Geneva: IPCC.

———. 2001. *Climate Change 2001: Mitigation, Impacts, Adaptation, and Vulnerability: Summaries for Policymakers.* Geneva: IPCC.

MacAlister Elliott and Partners-MEP. 1998. *Republic of Yemen: Fisheries Sector Review.* Yemen: European Commission.

O'Toole, M. J., L. V. Shannon, V. de Barros Neto, and D. E. Malan. 2001. "Integrated Management of the Benguela Current Region." In B. Von Bodungen and R. K. Turner, eds., *Science and Integrated Coastal Management.* Berlin: Dahlem University Press.

Sheppard, Charles, ed. 2000. *Seas at the Millennium: An Environmental Evaluation.* 3 vols. Oxford: Pergamon.

Sherman, Kenneth, ed. 1998. *Large Marine Ecosystems of the Indian Ocean: Assessment, Sustainability, and Management.* Oxford: Blackwell Science.

South African Department of Environmental Affairs and Tourism. 1999. *National State of the Environment Report.* Pretoria: DEAT.

———. 2000. *The South African Line Fishery: Past, Present, and Future.* Pretoria: DEAT.

Spalding, Mark D., Corinna Ravilious, and Edmund P. Green, in association with the UN Environment Programme and World Conservation Monitoring Centre. 2001. *World Atlas of Coral Reefs.* Berkeley: University of California Press.

UN Development Programme, UN Environment Programme, World Bank, and World Resources Institute. 2000. *World Resources 2000–2001: People and Ecosystems, The Fraying Web of Life.* Washington, DC: World Resources Institute.

UN Environment Programme. 1999. *Overview of Land-base Sources and Activities Affecting the Marine, Coastal, and Associated Freshwater Environments in the West and Central Africa Region.* UNEP Regional Seas Report and Studies No. 171. Nairobi: UNEP.

―――. 2002a. *Africa Environment Outlook*. Hertfordshire, UK: Earthprint Limited and UNEP.

―――. 2002b. *GEO: Global Environment Outlook 3*. London: Earthscan.

UN Food and Agriculture Organization. 1996a. "Fisheries and Aquaculture in Sub-Saharan Africa: Situation and Outlook in 1996." FAO No. 922 FIPP/C922. Rome: FAO.

―――. 1996b. "Fisheries and Aquaculture in the Near East and North Africa: Situation and Outlook in 1996." FAO No. 919 FIPP/C919. Rome: FAO.

―――. 2001. *FAOSTAT Online Statistic Service*. Rome: FAO.

―――. 2002. *The State of World Fisheries and Aquaculture*. Rome: FAO.

UN Industrial Development Organization. "Water Pollution Control and Biodiversity Conservation." UNIDO Media Corner. Available at http://www.unido.org/doc/100452.htmls (accessed December 7, 2002).

Wilkinson, Clive, ed. 2000. *Status of Coral Reefs of the World*. Townsville: Australian Institute of Marine Science and Global Coral Reef Monitoring Network.

World Bank. 2001. *African Development Indicators 2001*. Washington, DC: World Bank.

World Resources Institute. 2000. *World Resources 2000–2001, People and Ecosystems: The Fraying Web of Life*. Washington, DC: World Resources Institute.

World Wide Fund for Nature. "Africa Programme: Oceans and Coasts." Available at http://www.panda.org/africa/oceans.htm (accessed November 15, 2002).

8

Energy and Transportation

Energy use and transportation patterns in both Africa and the Middle East have come under increased scrutiny in recent years, driven by rising internal and international concerns about environmental trends across the regions. In Africa, energy consumption is steadily rising, fueled by growing populations and expanding motor vehicle fleets. Today, the transportation sector is the single greatest factor in the continent's steadily growing appetite for environmentally unfriendly fossil fuels, though high-polluting coal-fired power plants and industrial operations also deliver punishing blows to Africa's air, water, and land. Recognition of environmental problems associated with current practices is growing, but implementation of clean technology and other reform efforts in the energy and transportation sectors is complicated by endemic poverty, gaping holes in institutional infrastructure, and governmental apathy. Similar trends in fossil fuel consumption are also evident in the Middle East's energy and transport sectors, prompting growing concern about the future health of land- and marine-based ecosystems. In addition, the Middle East's economic fortunes remain closely tied to global oil consumption that has been criticized for damaging human health, compromising the ecological integrity of natural habitats, and contributing to global climate change.

Fossil Fuels and Their Environmental Impact

The environmental impact of fossil fuel dependence—locating, extracting, transporting, and consuming oil, coal, and natural gas—has drawn fire from conservation groups, health organizations, and policy-makers all over the world in the past half-century. By most measurements, antipathy is greatest toward coal, which takes the greatest toll on public health and ecosystem integrity of all the fossil fuels. Coal extraction and transport operations are notorious for degrading and eroding soils, destroying or fragmenting species-rich habitats such as forests and meadowlands, and contaminating watersheds (through wastewater discharges, toxic tailings, and landscape changes). At the consumption end, meanwhile, the consequences are even more far reaching. Burning coal produces heavy emissions of air pollutants, acid rain, and carbon dioxide, the main heat-trapping "greenhouse gas" responsible for encroaching global climate change. In addition, wastewater generated from coal-based thermal power generation contains residual chlorine, chromium/zinc sulfates, and other dissolved and suspended solids. Moreover, wastewater is generally of a higher temperature than the streams and rivers into which it is deposited, which alters downstream feeding, spawning, and other aspects of aquatic ecosystems.

Oil and natural gas extraction and delivery operations have also come under fire for their impact on the world's air, land, and water resources. Critics in the environmental and scientific communities contend that oil and gas development operations—which include roadways, pipelines, and other infrastructure—slice up important breeding and migratory areas, contaminate fragile rivers and aquifers with industrial pollutants, and diminish the wilderness character of undeveloped areas. And while natural gas consumption does not degrade surrounding air or water, emissions from oil consumption are a major contributor to the world's struggles with air pollution and global warming.

It is widely recognized that the harmful effects associated with fossil fuel extraction and use can be especially pronounced in places such as Africa, where government ownership of utilities and natural resources is the rule and the economic and political power of ordinary citizens—who are most directly affected by environmentally destructive operations—is often severely limited. The often grim situation is further exacerbated by funding limitations and official corruption that make it difficult for countries to implement pollution-mitigation technologies that have enabled developed countries at least to diminish the ecological impact of their energy appetites. Indeed, air, water, and land pollution issues often receive relatively low priority. One reason for this is institutional reluctance to interfere with industries that have taken root, generating jobs and money for communities. "African countries act to protect the somewhat limited, but hard-won, gains in industrialization, because these fledgling industrial sectors are considered essential to further economic and social development," confirmed the UN Environment Programme. "Any abatement strategies which burden them in any way, or which increase their financial and operational risks, are considered unacceptable" (UN Environment Programme 2002a).

But other initiatives that might relieve environmental problems associated with fossil fuel extraction and consumption are torpedoed by other pressing social concerns or government indifference to the plight of communities and stressed animal and plant species (Rosenblum and Williamson 1987; Ayittey 1998; UN Environment Programme 2002a). Indeed, "it is generally recognized by western operators that environmental standards in the African oil industry have been lacking. Most African governments in the past did not have the desire or the will to impose stringent checks or legal requirements, while the oil companies—like all commercial operators—rarely do more than is asked of them" (Ford 2001).

International nongovernmental organizations (NGOs) in the environmental and human rights fields charge that oil, natural gas, and coal exploration and extraction activities have taken a particularly devastating toll on rural communities and indigenous groups in Africa. By the time some of these extraction activities conclude, local people are left with the wreckage of once-vibrant forest and river ecosystems. These wounds are further aggravated by the fact that rural communities often receive little financial compensation that would enable them to build infrastructure (schools, hospitals, electricity, running water) or make other improvements.

Examples of this scenario are depressingly numerous and run from one end of the continent to the other. Possibly the most notorious case is Nigeria, where the state has exported more than $320 billion of crude oil since the 1970s "but has little to show for it beyond some decaying freeways and sports stadiums" (Robinson 2002). Many of these oil extraction operations have been concentrated in the Niger Delta, a treasurehouse of African biodiversity. But the indigenous Ogoni people who make their home in this region have received meager compensation from the government or the oil companies for the invasion, which by many accounts has despoiled large tracts of forests and wetlands. "Despite the vast wealth produced from the oil found under the Delta, the region remains poorer than the national average; and . . . the divisions between the rich and poor are more obvious in the areas where gas flares light up the night sky" (Human Rights Watch 1999).

As industrial operations in the Niger Delta devastated Ogoni land and fisheries, scattered protests coalesced into a movement that received international attention. "Oil exploration has turned Ogoni into a wasteland," charged Ogoni environmentalist and civil rights activist Ken Saro-Wiwa. "In return we have received nothing. . . . This action has destroyed wildlife and plant life, and has made the residents half-deaf and prone to respiratory diseases" (Saro-Wiwa 1992). But the protests sparked a violent crackdown by Nigerian authorities, and in 1995 Saro-Wiwa was executed despite an international outcry.

Today, the situation in the Niger Delta remains a source of major controversy, though international media coverage has faded. Multinational oil companies operating in the delta contend that they are a positive force in the region, funding infrastructure development and respecting indigenous rights. Critics, though, contend that they and the military rulers of Nigeria continue to destroy the lives and livelihoods of nearby communities. "Everywhere we visited we witnessed the destruction of the local environment, and the oppression of communities affected by what can accurately be described as an outlaw oil industry," charged one report. "Over the last 40 years, billions of dollars in profits are earned each year, as millions of barrels of oil are extracted. Meanwhile, high unemployment, failing crops, declining wild fisheries, poisoned waters, dying forests and vanishing wildlife are draining the very life blood of the region. Even the rainwater is acidic and poisoned" (Essential Action 2000).

The situation in the Niger Delta and some other corners of Africa has prompted some observers to label a number of African governments as "vampire states" eager to liquidate their natural resources for short-term financial gain, without regard for the burdens that these extractive activities impose on their people and their quality of life (Ayittey 1998). According to environmental activist Wangari Maathai, whose Green Belt Movement is one of the world's greatest grassroots success stories, "the tragic truth is that much of the [African] continent is being impoverished by greedy and egocentric leaders assisted by international companies who take advantage of the fact that some presidents run their country as if it were their personal property. Oppressed, cowed, and living in debilitating poverty, the majority of Africans can only watch as their leaders mortgage them and their lands with projects they neither need nor want" (Maathai 2003). But of course, this dynamic is not limited to Africa. "The Ogoni people of Nigeria whom Ken Saro-Wiwa represented, are just one of hundreds of marginalized communities around the world who are losing their livelihoods, traditional cultures and even their lives as loggers, ranchers, and oil drillers cash in on their environments," stated one analysis. "And all too often, as the Nigerian tragedy demonstrates, when people organize to defend themselves and to request compensation for lost jobs and deteriorating health, their appeals are met with harassment, beatings, imprisonment, or even murder" (Sachs 1995).

Overview of Energy Resources and Consumption
Oil: Lifeblood of the Middle East, Emerging Force in Africa
The Middle East has long been known as the planet's leading storehouse of oil, which remains the world's foremost fossil fuel for transportation and power

Petroleum tank farm in Dahran. COREL

generation. The nations of the Middle East—or West Asia, as it is sometimes known—contain more than 65 percent of the world's proved oil reserves, with 685.6 billion barrels of the world's known reserves of 1,050 billion barrels. By contrast, the region with the next highest total of proved reserves is South and Central America at 96 billion barrels (9.1 percent of the global total). Africa contributes another 76.7 billion barrels—7.3 percent of the global total— though speculation is rampant that the western side of the continent will be the next great oil exploration frontier. In terms of consumption, meanwhile, both Africa and the Middle East are minor players, albeit for different reasons. Africa is home to 13 percent of the world population, but it is saddled with socioeconomic problems that drag down consumption of energy. As a result of this impoverishment, Africa is responsible for only 3.3 percent of global oil consumption, at 2,490 thousand barrels daily (BP 2002).

This rate of oil consumption corresponds roughly with Africa's overall energy consumption; it consumes only 3 percent of total world energy produced, and its per capita energy consumption is only 23 percent of the global average. This low level of energy consumption occurs even though many nations are inefficient users of energy. For example, in terms of per unit industrial output, South Africa's paper manufacturers consume 30 percent more energy than their Western European counterparts, and South African steel manufacturers use over 50 percent more energy than European steel producers (Scholand

Figure 8.1 Global Proved Oil Reserves, 2001 (Thousand million barrels)

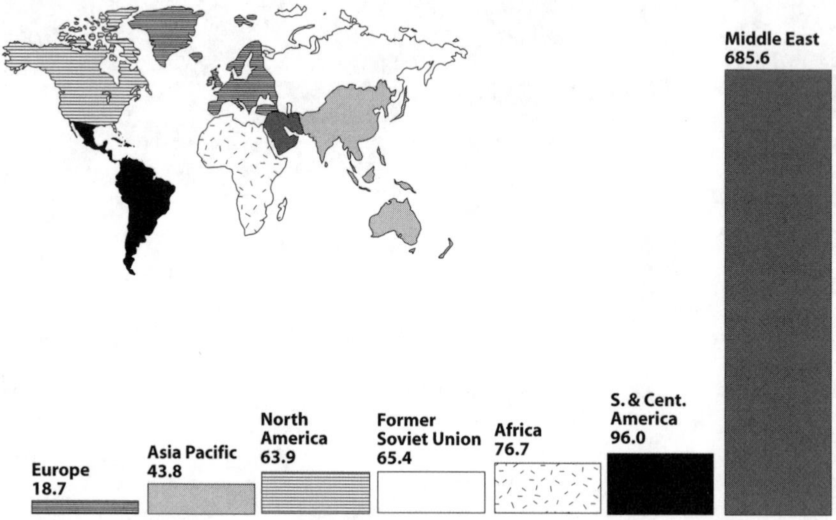

Middle East
685.6

S. & Cent.
America
96.0

Africa
76.7

Former
Soviet Union
65.4

North
America
63.9

Asia Pacific
43.8

Europe
18.7

SOURCE: BP. 2002. *BP Statistical Review of World Energy June 2002*. p. 5.
http://www.bp.com/files/10/statistical_review_1087.pdf. Accessed July 22, 2003.

1996). But despite such wastefulness, overall energy use on the continent is low because nearly 65 percent of African communities do not even have regular access to commercial forms of energy such as gasoline, coal, natural gas, or electricity. In addition, fewer than one in ten Africans have access to electricity. In fact, in 1997, five nations—South Africa, Egypt, Algeria, Nigeria, and Libya—accounted for 78 percent of all energy consumption and 84 percent of all energy production on the continent (Nevin 2002). Four of the above-mentioned countries are located in North Africa, which has experienced a much greater surge in energy consumption—jumping by 50 percent overall during the past two decades—than the rest of the continent (Organization of Arab Petroleum Exporting Countries 2002).

The Middle East, on the other hand, is also a modest consumer of oil, accounting for 4.8 percent of the global total at 4,306 thousand barrels daily. This relatively low share is attributable not to poverty but rather to its comparatively small overall population—though that too is on the rise (BP 2002). Indeed, on a per capita basis, oil consumption in the Middle East is closer to that of the industrialized world than that of most African countries, in part because big oil-producing states subsidize domestic oil consumption in ways that provide no incentive for conservation. In addition, the Middle East stamps a massive ecological "footprint" on the rest of the world via its huge oil exports to the United States and elsewhere.

Historically, the presence of vast oil fields in the Middle East has given the region little incentive to explore nuclear energy, a technology in which other parts of the world such as Europe and the United States invested heavily in the 1970s and 1980s. Indeed, no nuclear power facilities exist in the Middle East. In Africa, South Africa is the only nation with nuclear power, and it accounts for only 0.4 percent of global nuclear energy use (ibid.).

Global dependence on Middle East oil is expected to be strong for the next several decades. Currently the source of about 35 percent of the world's oil supply, it is expected to deliver more than 40 percent of the world's supply by 2015 (Townsend 2001). This continued demand, which is key to the overall economic stability of the region, is expected despite mounting concerns about possible supply disruptions caused by armed conflict and civil unrest. In fact, the troubled political landscape that endures in a number of these countries—including repressive governments, tensions between neighboring states (especially between Israel and Arab neighbors), and religious extremism that has bred severe anti-Western attitudes—has dampened

The Chad-Cameroon Oil Pipeline

One of the most significant and controversial energy projects to be undertaken in Africa in recent years is the World Bank–sponsored Chad-Cameroon Petroleum Development and Pipeline Project. This initiative consists of three distinct but interrelated projects: development of oil fields at Doba in southern Chad; creation of offshore oil-loading facilities on Cameroon's Atlantic coast; and construction of a 1,070-kilometer (660-mile) pipeline that will connect the two. But it is the pipeline—construction of which is expected to account for U.S.$2.2 billion of the project's total U.S.$4 billion cost—that has emerged as the focal point of a fierce international debate. Supporters of the pipeline assert that it will be an economic godsend for the impoverished people of Chad and Cameroon. Opponents are concerned about the project's potential harm to fragile ecosystems and indigenous human communities.

The pipeline, which is being developed by a consortium of international oil companies and the governments of Chad and Cameroon—the latter through loans from the World Bank—may prove enormously beneficial to the two West African nations. According to the World Bank, the project could generate nearly U.S.$2 billion in revenues for Chad and U.S.$500 million for Cameroon over its twenty-five-year production period. "This project could transform the economy of Chad. The country is so poor at present that it cannot afford the minimum public services necessary for

(continues)

a decent life. By 2004, the pipeline would increase Government revenues by 45 to 50 percent per year and allow it to use those resources for important investments in health, education, environment, infrastructure, and rural development, necessary to reduce poverty" (World Bank 2002).

Already, the pipeline project has emerged as a major source of jobs for people in the two nations, who struggle with very high rates of unemployment. In late 2002 more than 12,000 workers were employed in the project's pipeline, oil field, and marine development phases, with 80 percent of employees hailing from Chad and Cameroon.

However, tangible gains in employment and the promise of increased revenue for the region have failed to sway detractors. Organized opposition to the Chad-Cameroon pipeline coalesced in the late 1990s. By that time, neighboring Nigeria had built a notorious record of environmental and human rights abuses in its own pursuit of oil riches, and local, national, and international NGOs feared that the Chad-Cameroon pipeline project had the potential to wreak similar havoc.

One of the chief objections raised by the NGOs—which included religious groups, human rights organizations including Amnesty International, and environmental groups such as Friends of the Earth and West Africa–based Environmental Rights Action—was that the project did not provide adequate oversight for distribution of the revenue generated by the pipeline. Critics charge that official corruption is rife in both Cameroon and Chad, and that "profits from the Chad/Cameroon oil project will go largely unaccounted for, despite the fact that both countries are desperately in need of poverty

(continues)

Western enthusiasm for oil production investment in the region in recent years. Moreover, some states have placed strict limits on investment and operational participation by foreign companies, a reflection of concerns about energy security and public opinion (International Energy Agency 2002). For example, Saudi Arabia has made accommodations to welcome international firms into new natural gas development schemes, but all of its oil operations—the cornerstone of its entire socioeconomic structure—remain strictly off limits to foreign companies.

Of all the oil-producing states in the Middle East, Saudi Arabia has the most to gain from continued international dependence on oil. It contains one-quarter of the world's proven oil reserves (264.2 billion barrels), and in 2002 it produced about 8 million barrels a day. Oil revenues constitute 90 to 95 percent of total Saudi export earnings, 70 to 80 percent of state revenues, and around 40 percent of the country's gross domestic product, despite re-

alleviation, health care, education, and environmental protection programs" (Rainforest Action Network, n.d.).

Concern was also expressed about the possible impact of the pipeline on rural and indigenous peoples, from disruption of traditional lifestyles to outright displacement. For example, many indigenous Bakola people, commonly known as Pygmies, resided in areas targeted for pipeline development.

Finally, objections were raised on environmental grounds. Opponents of the planned pipeline noted that a spill could pollute rivers that provide drinking water for millions of people in the two countries. They also noted that the pipeline would be carved through two major rain forests in Cameroon— the inland Deng Deng Forest and forests along the Kribi Coast—that contain numerous endangered animal and plant species, including gorillas, forest elephants, and black rhinos.

Resistance to the plan became so strong that two oil companies dropped out of the scheme (though they were replaced), and the World Bank called for a new round of impact studies. New environmental and financial safeguards were subsequently implemented, and these measures have elicited measured praise from some members of the environmental community. These changes include anticorrosion protection for the pipeline, full payment of relocation costs for affected people and communities, and long-term decommissioning and reclamation plans (Ford 2001).

Perhaps most significantly, the campaign prompted pipeline planners to make major changes in the proposed route. For example, planners shifted the pipeline route out of the ecologically

(continues)

peated attempts to introduce a greater level of diversification into the economy. More than half of these reserves are contained in only eight of its eighty oil and gas fields; of these, the most notable operations are Ghawar, the world's largest land-based oil field, with estimated remaining reserves of 70 billion barrels, and Safaniya, the world's largest offshore oilfield, with estimated reserves of 19 billion barrels (Energy Information Agency 2002).

Iraq has the second largest oil reserves in the region—and the world—at over 112 billion barrels of known oil, about 10 percent of the global total. But sanctions imposed by the international community in the wake of the 1991 Persian Gulf War have stifled the country's production capacity and restricted its ability to export oil. Moreover, future ownership and use of Iraq's huge oil deposits is murky. In 2003 the United States toppled the regime of Iraqi dictator Saddam Hussein, but the identity and character of the government that will ultimately take the country's reins remains unknown.

fragile Mbere Rift Valley to a plateau high above the valley. The heart of the Deng Deng Forest was spared as well, as engineers moved the route away from the untouched interior in favor of a route that parallels an already existing railroad line. And the Kribi rain forest on Cameroon's Atlantic coast received similar clemency, as the route was changed so that it shadowed an existing roadway instead of slicing through virgin forest (ibid.). Finally, the oil companies investing in the project agreed to donate U.S.$2.9 million to help create two national parks in Cameroon (World Bank 2002).

Despite these improvements, the pipeline project still concerns some environmentalists and social justice activists. For example, the consortium has budgeted only $800,000 for a major oil spill disaster, even though such an event could cause millions of dollars of environmental damage (Ford 2001). And worries about the ultimate disposition of pipeline revenue linger, given the poor track record of the region's political leadership in such matters.

As debate about the impact of the pipeline continues, the project itself is nearing completion. As of September 2002 approximately two-thirds of the pipeline had been completed, and planners said that the pipeline was advancing another 3 kilometers deeper into the forest every day on its journey to Cameroon's Atlantic Ocean shores.

Sources:

Ford, Neil. 2001. "Pollution Worries Mount." *African Business* 261 (January).

Rainforest Action Network. "Case Study: Chad-Cameroon Pipeline." n.d. Available at http://www.ran.org/ran_campaigns/citigroup/cs_chadcam.html (accessed February 2003).

World Bank. 2002. "The Chad-Cameroon Petroleum Development and Pipeline Project." December 18. Available at http://www.worldbank.org/afr/ccproj/project/pro_overview.htm (accessed February 2003).

The political situation in the Middle East—combined with ever-expanding global appetites for oil—has also contributed to a surge in interest in oil exploration in West and Central Africa. Investment in this region has surged in the last two decades, and some industry experts have even forecast that Africa might eventually eclipse the Middle East as the planet's leading exporter of oil. Already, expanded oil production activity has sparked significant social, economic, and environmental changes in some targeted areas. For example, Malabo, the capital of tiny Equatorial Guinea, has experienced significant economic growth in recent years as a result of a series of major offshore oil operations. Production of high-quality crude oil from these operations rose from 17,000 barrels a day in 1996 to more than 220,000 barrels a day in 2002, pushing annual per capita income up by 150 percent during that time span

(Robinson 2002). "The bonanza in Equatorial Guinea is being repeated across the region," observed one report. "Chad, one of the world's poorest countries, will soon start pumping more than 200,000 barrels a day through a $3.7 billion, 660-mile pipeline—one of Africa's biggest-ever infrastructure projects—that crosses Cameroon. The island nation of São Tomé and Príncipe, which sits on perhaps 4 billion barrels of crude, is also attracting foreign oilmen" (ibid.). In addition, established oil-exporting nations in the region such as Nigeria, Angola, and Gabon are actively seeking to increase their output. In these and other African countries along the Pacific Coast, such as Ivory Coast and Morocco, offshore exploration has been a primary focus (International Energy Agency 2002).

Deep sea oil exploration off the shores of West and Central Africa is expanding primarily because of the perceived volume and quality of the reserves (the oil's low sulfur content makes it easy to refine, and the location of fields is attractive for transport to major European and North American markets). But emphasis on offshore sites is also attributable in part to reluctance to invest in land-based operations that are more vulnerable to political unrest and armed conflict, both of which are rife in parts of West Africa. For example, oil exploration and production continues to proceed without incident in the placid waters off Angola's coastline, isolated from the civil war that rages across that land (Center for Strategic and International Studies 2000).

Advocates of these rapidly expanding offshore operations contend that they can provide impoverished West African nations with badly needed revenue to improve education, invest in hospitals and medical care, feed malnourished families, and institute environmental protections for marine and land-based resources alike. In places like the above-mentioned Malabo, at least some of these benefits seem to be taking place. But skepticism about the allocation of oil revenue in Angola and other countries that are politically unstable or saddled with corrupt leadership is high. "If the past is prologue, then little of the coming oil wealth will pass to the benefit of the Angolan people," predicted one analysis. "Rather, the hard currency earned from oil exports more likely will be employed in support of efforts by the [government] to subdue all opposition, and that, in turn, can only translate into further devastation" (ibid.).

Moreover, marine extraction and transport of oil is controversial on environmental grounds. Proponents of offshore drilling and transport contend that operations are environmentally safe; they charge that environmentalists exaggerate the dangers of oil and gas exploration and extraction activities. But oil spills have soiled the marine waters of both Africa and the Middle East in recent years, with incidents reported in Africa's Cape of Good Hope, West

Africa's Gulf of Guinea, the Arabian Gulf (Persian Gulf), the Arabian Sea, the Gulf of Aden, and the Gulf of Oman. In addition, oil contamination has also been traced to tankers that empty oil-laced ballast on the high seas; leaks from oil drilling activities and refineries located in coastal zones; and leaks from barges, tankers, fishing vessels, and other boats (UN Environment Programme 2002a). In addition, seabed mining for oil and other minerals has been criticized for disturbing coastal wetlands, estuaries, and sand dune systems; diminishing local air quality; and altering the habitat of fish and other marine species (South African Department of Environmental Affairs and Tourism 1999).

Oil Fuels Growing Transport Sector

Thanks in part to the ready availability of inexpensive gasoline in the Middle East, rates of motor vehicle use and private vehicle ownership are on the upswing across the region. Similar trends are evident throughout much of Africa as well. But although the expanding size of the motor vehicle fleet has been credited with making education, health care, and employment more accessible and with spurring economic development, there have also been pronounced negative consequences in the realms of environment and quality of life (Sperling and Salon 2002).

One of the major environmental drawbacks associated with the growing transport sector is loss or degradation of forests, wetlands, waterways and other natural areas to roads, pipelines, drilling facilities, refineries, and other infrastructure. These losses have a corresponding impact on the health and vitality of numerous species of flora and fauna. The other obvious disadvantage of increased motor vehicle use is increased emissions of chemical compounds (such as lead) that diminish air quality and contribute to the global warming phenomenon (such as carbon dioxide, the leading greenhouse gas). These problems are especially pronounced in the cities of Africa and the Middle East. "When the more developed countries were building their transportation infrastructure, their populations were small compared to those in much of today's developing world, and the cost of motorized vehicles was relatively high. Today's megacities of the developing world are already huge and still expanding. There is little time or money to build public transportation systems or to expand roads to handle the new traffic. They are already experiencing serious congestion, economic and environmental damage, and major safety problems" (ibid.).

Indeed, neither the Middle East nor Africa has had much success in controlling the shape and character of transportation. In most Middle East nations, growing numbers of privately owned automobiles and trucks clog city streets and roadways. Many of these vehicles are older models that are not equipped with modern pollution mitigation measures, so they spew much higher levels of emissions such as hydrocarbons and nitrogen oxides than do newer models. Their impact is further exacerbated by the widespread availability of inexpensive gasoline and the lack of investment in mass transit, both of which discourage a meaningful conservation ethic from taking root in the population.

Africa's transportation infrastructure, meanwhile, is in very poor shape. In many countries, congested urban areas and battered roads are knotted with high-polluting automobiles, trucks, motorcycles, and buses that are steadily swelling in number because of modest improvements in buying power. The continent's fleet remains dominated by older vehicles, however, that belch out high levels of pollutants—and that are not retired until every last mile of life has been wrung out of them. In addition, some of the cars and trucks that have been introduced to Africa's roadways in recent years are themselves older models that are not always equipped with modern pollution-abatement equipment. In northern Africa, for instance, the number of motor vehicles in most countries has nearly doubled in the past ten to fifteen years, but most of the cars and trucks are older models. These older vehicles emit twenty times more hydrocarbons and carbon monoxide and four times more nitrogen oxides than new vehicles. In addition, studies indicate that particulate emissions from poorly maintained diesel-fueled buses and trucks are five to seven times higher than those from similar but well-maintained vehicles. Finally, high taxes on the importation of new motor vehicles have further exacerbated Africa's reliance on older automobiles and trucks (Larson 1995; UN Environment Programme 2002a).

Emissions from motor vehicles constitute Africa's major source of lead contamination, for only a few countries have thus far made the transition to unleaded gasoline. This long-standing reliance on leaded fuel has been cited as a contributing factor to high rates of respiratory and nervous-system problems, especially among children subject to high levels of exposure—primarily those living in Africa's expanding urban centers. Fortunately, leaded gasoline seems destined to become a relic of the past in Africa within a decade or so. In 2003 the UN Environment Programme declared that Africa was accelerating its switch toward unleaded gasoline, and it forecast that leaded fuel would be a rarity in most African countries by 2010. Four African countries—Egypt, Libya, Mauritius, and Sudan—are already fully lead-free, and countries such as Morocco and Tunisia are on the verge of full compliance. Another twenty countries, including Eritrea, Ghana, Kenya, Nigeria, South Africa, Togo, and Uganda, have drawn up plans—or were in the process of making plans—to phase out leaded gasoline by 2005 or 2006.

Vehicle fleets are growing in Ghana and other African countries. LIZ GILBERT/CORBIS

Another major impediment to reversing downward trends in air quality in Africa's cities is continued reliance on diesel gasoline. Diesel fuel is the preferred choice for commercial trucks and public buses in many parts of Africa because of its greater fuel efficiency and lower cost. But Africa trails other parts of the world in introducing technological innovations that reduce the environmental toll of diesel use. In Africa, emissions from diesel consumption continue to include high levels of cancer-causing agents such as arsenic, benzene, nickel, and formaldehyde. Finally, the transportation sector's consumption of oil could be relieved somewhat if public transit systems were more widespread or attractive. But investments in this realm have historically ranged from modest to nonexistent, as few countries have been able to afford the expense associated with such projects.

Natural Gas—Another Source of Resource Wealth

The Middle East and the former Soviet Union rank as the regions with the greatest abundance of known natural gas reserves. The former Soviet Union holds the nominal crown in this regard, with 36.2 percent of the world's proved natural gas reserves. But the Middle East is one major discovery away from overtaking Russia, for it holds 36.1 percent of the globe's known reserves (55.91 trillion cubic meters/1974.6 trillion cubic feet). Moreover, the Middle East houses four of the top five countries in the world in terms of known natural gas holdings, with Iran, Qatar, Saudi Arabia, and the United Arab Emirates holding the second through fifth positions behind Russia. Africa, meanwhile, contains 7.2 percent of the world's proved natural gas reserves (11.18 trillion cubic meters/394.8 trillion cubic feet), though geographic distribution of these fields is uneven across the continent. In terms of consumption of natural gas, the countries of the Middle East used 201.5 billion cubic meters in 2001, approximately 8.4 percent of the global total. In addition, the region's level of natural gas consumption is forecast to rise considerably in the next several years, perhaps doubling in the first two decades of the twenty-first century. Africa, meanwhile, is only a minor factor in global natural gas use. In 2001 it consumed 60.2 billion cubic meters of natural gas, about 2.5 percent of the worldwide total (BP 2002).

The Middle East is thus poised to take advantage of the growing global consensus that natural gas is the best and most environmentally friendly of the fossil fuels used for power generation. In addition to its known reserves, it is actively seeking out new deposits to tap for domestic use. African consumption of natural gas may also increase in the near future, especially if states in the western and central reaches of the continent follow through on their stated interest in harnessing natural gas that is currently flared off in oil

Figure 8.2 Global Proved Natural Reserves, 2001 (Trillion cubic meters)

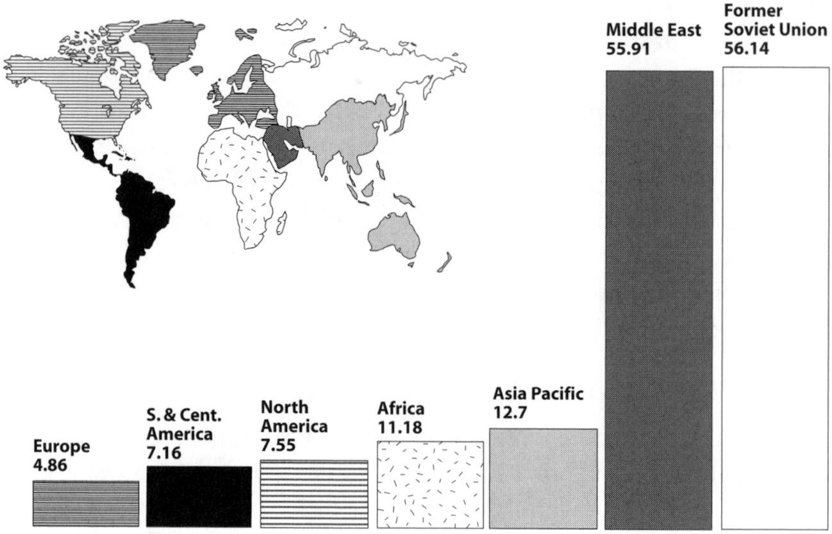

SOURCE: BP. 2002. BP Statistical Review of World Energy June 2002. p. 21.
http://www.bp.com/files/10/statistical_review_1087.pdf. Accessed July 22, 2003.

production. For example, in the late 1990s a report funded by the World Conservation Union-IUCN reported that fully 75 percent of Nigerian natural gas is flared, far exceeding any other nation's allowable flaring limits (Ashton-Jones et al. 1998). This flaring activity is harmful to human health and ecosystems and also contributes to global warming. Indeed, gas flaring is responsible for approximately 1 percent of global emissions of carbon dioxide, the main greenhouse gas implicated in climate change. Fortunately, some African nations are responding to this issue, albeit belatedly. Angola, for example, has prohibited the flaring of natural gas in all new production contracts with the oil industry (Ford 2001).

Africa's biggest project in natural gas development is the proposed West Africa Gas Pipeline Project, which seeks to connect gas-rich Nigeria to markets in Benin, Ghana, and Togo. The environmental benefits and drawbacks of this pipeline, which will run 620 kilometers when completed, have been fiercely debated. Advocates say that it will increase energy efficiency in target markets and reduce dependence on more environmentally destructive fossil fuels. Detractors claim that the pipeline could have adverse impacts on rivers and forests in the region.

Coal gasification plant in South Africa. JONATHAN GLAIR/CORBIS

Reliance on Coal High in Some Areas

Africa and the Middle East are not major players in the world of coal production. Together, they contain only 5.8 percent of the planet's known coal reserves, about 57 billion tons. Neither region is a heavy consumer of coal, either. Together, they account for only about 4.3 percent of worldwide consumption (BP 2002). A notable exception in this regard, however, is South Africa, the leading economy of sub-Saharan Africa. The land on which the South African people live and work is riddled with extensive veins of coal, which became the major fuel for the country's economic engine during the twentieth century. Today, South Africa receives 90 percent of its electricity from coal-fired power stations, making it more dependent on coal—the most environmentally damaging of all fossil fuels—than any other nation except China (Scholand 1996).

Coal's central role in power generation in South Africa—sub-Saharan Africa's leading economy—is troubling on several fronts. Coal use is a public health concern because it is a major source of atmospheric sulfur, nitrogen oxides, and greenhouse gases—especially carbon dioxide—when burned, and coal mining has been criticized for polluting vital rivers and aquifers, disrupting watershed drainage patterns, destroying forests and other species-rich habitat, and ruining the scenic quality of natural areas. In fact, detractors contend that coal wreaks a heavy toll on human health and natural ecosystems at every stage of development, from extraction through transport to consumption (UN Environment Programme, Collaborating Centre on Energy and Environment 1997). Moreover, energy derived from coal consumption continues to be enjoyed primarily by white, urban populations over black, rural communities, even in this postapartheid era. For instance, in the mid-1990s electricity distributed by South Africa's government-owned national electric utility reached 98 percent of white households but only 31 percent of black households (and only 20 percent of black households in rural communities). "The coal-rich province of Mpumalanga, east of Johannesburg and Pretoria, contains some of the largest power plants in the world," observed one analysis. "Residents of Johannesburg get the electricity; local people get an overdose of air pollution, acid rain, and other forms of environmental damage" (Scholand 1996).

In southern Africa's Zambezi basin—a watershed that drains approximately 500,000 square miles of the continent—mining operations targeting deposits of coal and other minerals including tin, copper, and gold have been blamed for widespread water acidification and discharges of highly toxic metals such as arsenic, mercury, cadmium, and lead. In places like South Africa, Mozambique, and Zimbabwe, both open and underground mining activities proceed with only modest environmental protections, and these extraction processes have been faulted for cutting into previously undeveloped natural areas and generating huge amounts of mining waste. The Kafue River corridor, for example, has more than 90,000 hectares of land under mining, 11 percent of which is occupied by solid waste dumps. The solid waste, which in 1996–1997 amounted to 17 million tons, includes overburden, waste rock, tailings, and smelting slag tainted with various toxic heavy metals (Chenje 2000).

These insults to the land are possible only because most countries have made little effort to draft or implement laws protecting their natural resources. In Zimbabwe, for instance, the Mines and Minerals Act supersedes most other acts, placing few restrictions on the exploitation of mining rights once the mining permit has been obtained. "Thus, the Act does not prevent extensive tree cutting without reforestation, poaching by mine workers, siltation [of rivers], dumps and noncompliance with quittance requirements when mines are closed" (UN Environment Programme, Collaborating Centre on Energy and Environment 1997).

Hydroelectric Power and
Other Forms of Renewable Energy

In Africa, hydroelectric power constitutes an important part of the overall energy grid, accounting for more than 80 percent of the total power in eighteen

African countries and more than 50 percent in twenty-five countries (World Commission on Dams 2001). These facilities are especially prevalent in the continent's eastern and western regions, where major river systems are located. In Egypt, for example, the Aswan High Dam's 2.1 million kilowatts of electricity generation capacity has successfully electrified the nation. But total electricity generation via hydropower in Africa is still dwarfed by that of other parts of the world, such as North America, Latin America, and Europe, each of which accounts for more than 20 percent of global hydroelectric consumption (BP 2002).

Hydroelectricity has been praised for many years as an energy resource with the capacity to generate significant amounts of energy without polluting the air with greenhouse gases and other pollutants. In addition, some dams have been built both to generate power and provide water for irrigation, a dual mandate that has boosted economic development while simultaneously enhancing agricultural activity. But potential growth of this energy source is limited in Africa. Most African states have little money available for such capital-intensive projects. In addition, many sites suitable for large hydroelectric installations have already been developed, though some significant river systems in southern and central Africa may yet to be harnessed. In the Middle East, meanwhile, hydropower has never been a large element of the energy picture because of the small number of major rivers in the region, though irrigation-oriented dams are commonplace.

Moreover, hydroelectric dams are no longer seen by the world community as an entirely benign source of energy. Concerns about dams' disruption of fragile freshwater fisheries and their inundation of important wildlife habitat and indigenous lands have soared in recent decades, and while dam-reservoir networks will undoubtedly still serve important functions in Africa and the Middle East for decades to come, assessments of their total environmental impact are emerging as a higher priority.

Other Forms of Renewable Energy: Poised for Growth?

Africa and the Middle East have both been cited as areas of high potential for solar and wind energy initiatives, because of their primarily sunny climates and the presence of windy and remote land-based and offshore regions. Some nations have actively pursued alternative energy development, both on their own and through bilateral and multilateral arrangements. In recent years, for example, Egypt and Israel have worked together on Noor Al Salam (Light of Peace), a World Bank–sponsored feasibility study to develop a major new natural gas/solar power plant. And in 2002, South Africa, which has long supported its electric infrastructure on a steady diet of high-polluting coal, declared its intention to supplement its coal energy with "green" energy. Among the measures announced by South Africa: a goal to meet 2 percent of national energy demand through renewables by 2012; establishment of a regulatory framework for renewable energy; and active solicitation of foreign investment in green power.

The potential of alternative sources of energy—solar, wind, small-scale hydropower, and biomass—may be particularly relevant for nations with large rural populations. Providing electricity to these remote communities is prohibitively expensive for many African states. But advocates of renewable energy believe that incorporating green technology into rural development planning can be both cost-effective and environmentally beneficial. In South Africa, for instance, "efficient solar homes would radically reduce indoor pollution by cutting the amount of heating fuel a family requires. And the savings on fuel would mean a substantial increase in disposable income for a large portion of the population. . . .. [For many South Africans], solar housing could cut deeply enough into their fuel expenses to increase disposable income by nearly 50 percent" (Scholand 1996).

Wood and Other Traditional Fuels
Remain Essential to African Households

Wood, charcoal, crop residues, animal dung, and other traditional biomass fuels remain a staple in many African households, both in rural areas and in the squatter settlements that ring urban centers. According to the International Energy Agency, biomass accounts for about 30 percent of the total energy supply in developing countries, with wood accounting for more than half of this fuel source. But in some nations in Africa and elsewhere, wood accounts for a much higher percentage of overall biomass consumption. In Uganda, Rwanda, and Tanzania in sub-Saharan Africa, for example, wood-fuels provide 80 percent or more of total energy requirements (Matthews 2000). Indeed, Africa and Asia together account for more than 75 percent of global woodfuel consumed.

The consequences of this heavy reliance on traditional biomass fuels are not trivial, either for the environment or for public health. In the former regard, demand for woodfuel can place heavy localized pressure on species-rich forest habitat. Indeed, in Africa and Asia, "rising demand for fuelwood and charcoal is causing a halo of deforestation around many cities, towns, and roads" (ibid.). In terms of human health, meanwhile, regular dependence on these fuels for cooking and heating exposes children and women (who are primarily responsible for cooking and other household duties in African homes)

Figure 8.3 Global Woodfuel Production, 1998

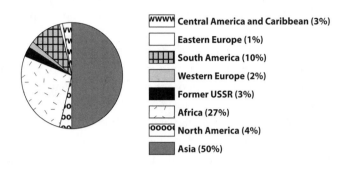

Central America and Caribbean (3%)

Eastern Europe (1%)

South America (10%)

Western Europe (2%)

Former USSR (3%)

Africa (27%)

North America (4%)

Asia (50%)

SOURCE: UN Food and Agriculture Organization. http://earthtrends.wri.org/text/FOR/features/ FOR_fea_woodfuel.htm. Accessed July 22, 2003.

to heightened concentrations of sulfur oxides, nitrogen oxides, carbon monoxide, and soot particles, all of which can produce respiratory problems. Evidence also suggests that as populations grow and encroach upon areas that previously provided sustainable amounts of biomass, shortfalls of woodfuel and other types of biomass are becoming more commonplace. "Numerous studies document instances of villagers traversing ever-longer distances to gather daily wood supplies. Woodfuel shortages are especially likely to occur near cities. . . . [But] there is good evidence that woodfuel supply in developing countries can be sustainable even in densely populated areas, where government planting programs, community woodlots, and plantations are adequately managed" (ibid.).

Addressing Energy-Related Environmental Problems in Africa and the Middle East

Recognition of energy-related environmental problems appears to be on the upswing in many states in both Africa and the Middle East. The long-term implications of fossil fuel dependency on economic development, political stability, public health, and environmental sustainability have all received heightened scrutiny in recent years. In some instances, this expanded sphere of considerations can be traced in large measure to pressure and prodding from other members of the international community, from multinational agencies to nongovernmental organizations. But some nations have exhibited a genuine commitment to change, exploring ways to introduce environmentally friendly renewable energy technologies or reduce the environmental impact of fossil fuel extraction and use. For example, in the United Arab Emirates, constituencies ranging from environmental groups to municipalities to oil industry representatives are all helping shape programs that place new emphasis on environmental protection and public health in oil exploration and production activities (Smith 1999).

If Africa and the Middle East hope to make emerging visions of environmental sustainability become a reality in their cities, villages, and natural areas, however, the nations that compose these regions will have to take bold and decisive steps to transform their societies and update their economic infrastructures. In Africa at least, such steps will be impossible without the assistance of other, more affluent members of the world community.

Experts agree that Africa and the Middle East must institute major reforms to their energy and transportation sectors. In the transport sector, these reforms should take a multitude of forms, including promotion of unleaded fuels and conversion to cleaner fuels for motor vehicles; removal or refitting of automobiles, trucks, motorcycles, and buses that emit high levels of pollution; increased emphasis on (and investment in) mass transit systems; increased fuel conservation through removal of gasoline subsidies; and improved urban planning to relieve congestion in major traffic corridors (UN Environment Programme 2002a; Sperling and Salon 2002; World Bank 2001).

Similarly, policy and investment changes in the larger energy sector must be made quickly to avoid potentially catastrophic economic, social, and environmental problems in the coming years. To accomplish this, a renewed sense of urgency—and opportunity—will be necessary. For example, in 1991 the Persian Gulf War deposited major oil spills in marine waters and fouled the air with great gouts of pollution from burning oil and gas wells. The war damaged large parts of the Gulf coastline, diminished fishery resources, contaminated water supplies, and hurt a variety of desert species of flora and fauna. In the aftermath of these dispiriting events, some governments in the region, such as Saudi Arabia, continued to let environmental protection languish. But other states, such as the United Arab Emirates and Oman, responded by stepping up their efforts to protect their environment (Smith 1999). Today, in this age of increasing ecological uncertainty, advocates of environmental sustainability contend that the nations of Africa and the Middle East need to exhibit a similar sense of obligation and responsibility to the natural world and its myriad inhabitants.

To this end, alternative energy sources should be developed and incorporated into state infrastructures—and over the space of years instead of decades—in order to reduce dependence on traditional fossil fuels implicated in global warming, habitat and species loss, and contamination of ocean and freshwater resources. Greater energy efficiency is also a pressing need in nearly every state in the region, as is investment in pollution mitigation technologies in industrial processes and power generation.

Many African and Middle Eastern nations also require basic improvements in political leadership, which too often maintain exclusionary political and socioeconomic structures. Indeed, "if all the vulnerable members of society— the impoverished, indigenous peoples, ethnic minorities, women, children— had access to environmental information and could exercise their right to free speech, then potential polluters and profligate consumers would no longer be able to treat them as expendable, and would have to seek alternatives to their polluting activities and their overconsumption" (Sachs 1995).

Sources:

Ashton-Jones, Nick, Susi Arnott, and Oronto Douglas. 1998. *The Human Ecosystems of the Niger Delta.* Port Harcourt, Nigeria: Environmental Rights Action.

Ayittey, George B. N. 1998. *Africa in Chaos.* New York: St. Martin's.

Bamber, Derek. 2001. "Bright Spots in a Cloudy Future." *Petroleum Economist* 68 (July).

Bhagavan, M. R., ed. 1999. *Reforming the Power Sector in Africa.* London, New York: Zed Books and African Energy Policy Research Network.

BP. 2002. *BP Statistical Review of World Energy 2002.* London: Group Media and Publications.

Carbon Dioxide Information Analysis Center. 2001. *Global, Regional, and National Fossil Fuel CO_2 Emissions.* Oak Ridge, TN: Oak Ridge National Library, DOE.

Center for Strategic and International Studies. 2000. *The Geopolitics of Energy into the 21st Century.* Washington, DC: CSIS.

Chenje, Munyaradzi, ed. 2000. *State of the Environment: Zambezi Basin.* Harare, Zimbabwe: SADC, IUCN, SARDC, ZRA.

Chenje, Munyaradzi, and Phyllis Johnson, eds. *State of the Environment in Southern Africa.* 1994. Harare, Zimbabwe: South African Research and Documentation Centre, IUCN-World Conservation Union, and Southern African Development Community.

Economides, Michael, and Ronald Oligney. 2000. *The Color of Oil: The History, the Money, and the Politics of the World's Biggest Business.* Katy, TX: Round Oak.

Energy Information Agency (U.S. Department of Energy). 2002. "Saudi Arabia." October. Available at http://www.eia.doe.gov/emeu/cabs/saudi.html (accessed February 2003).

Essential Action and Global Exchange. 2000. *Oil for Nothing: Multinational Corporations, Environmental Destruction, Death and Impunity in the Niger Delta.* January 25. Available at http://www.essentialaction.org/shell/report/index.html (accessed February 2003).

Ford, Neil. 2001. "Pollution Worries Mount." *African Business* 261 (January).

Hankins, Mark, and Robert J. Van der Plas. 1998. "Solar Electricity in Africa: A Reality." *Energy Policy* (March).

Held, Colbert C. 2000. *Middle East Patterns: Places, Peoples, and Politics.* Boulder, CO: Westview.

Human Rights Watch. 1999. *The Price of Oil: Corporate Responsibility and Human Rights Violations in Nigeria's Oil-Producing Communities.* New York: Human Rights Watch.

Intergovernmental Panel on Climate Change. 2001a. *Climate Change 2001: Mitigation, Impacts, Adaptation, and Vulnerability: Summaries for Policymakers.* Geneva: IPCC.

———. 2001b. *Climate Change 2001: The Scientific Basis.* Geneva: IPCC.

International Energy Agency. 2002. *World Energy Outlook 2002.* Paris: IEA.

Klare, Michael T. 2001. *Resource Wars: The New Landscape of Global Conflict.* New York: Metropolitan.

Koning, Henk W. De. 1990. "Air Pollution in Africa." *World Health* (January–February).

Larson, B. 1995. *Natural Resource Extraction, Pollution, Intensive Spending, and Inequities in the Middle East and North Africa.* Washington, DC: World Bank.

Maathai, Wangari. 2003. *The Green Belt Movement: Sharing the Approach and the Experience.* Rev. ed. New York: Lantern Books.

Matthews, Emily. 2000. "Undying Flame: The Continuing Demand for Wood as Fuel." In *Pilot Analysis of Global Ecosystems: Forest Ecosystems.* Washington, DC: World Resources Institute.

Nevin, Tom. 2002. "Africa's Energy Masterplan." *African Business* (December).

Organization of Arab Petroleum Exporting Countries. 2002. *Annual Statistical Report 2002.* Kuwait: OAPEC.

Roberts, Gwilym, and David Fowler. 1995. *Built by Oil.* Reading, Berkshire: Ithaca.

Robinson, Simon. 2002. "Black Gold: The U.S. Is Becoming Less Reliant on Middle Eastern Oil, Thanks to West Africa." *Time* 160 (December 23).

Rosenblum, Mort, and Doug Williamson. 1987. *Squandering Eden: Africa at the Edge.* San Diego: Harcourt, Brace, Jovanovich.

Sachs, Aaron. 1995. *Eco-Justice: Linking Human Rights and the Environment.* Worldwatch Paper 127. Washington, DC: Worldwatch.

Saro-Wiwa, Ken. 1992. *Genocide in Nigeria: The Ogoni Tragedy.* Port Harcourt, Nigeria: Saros International.

Scholand, Michael. 1996. "Re-Energizing South Africa." *World Watch* 9 (September–October).

Smith, Pamela Ann. 1999. "Protecting the Arab Environment." *Middle East* 286 (January).

South African Department of Environmental Affairs and Tourism. 1999. *National State of the Environment Report.* Pretoria: DEAT.

Sperling, Daniel, and Deborah Salon. 2002. *Transportation in Developing Countries: An Overview of Greenhouse Gas Reduction Strategies.* Prepared for the Pew Center on Global Climate Change. Washington, DC: May.

Townsend, David. 2001. "The Balance of Power." *Petroleum Economist* 68 (November).

UN Development Programme. 2000. *World Energy Assessment: Energy and the Challenge of Sustainability.* New York: UNDP.

UN Development Programme, UN Environment Programme, World Bank, and World Resources Institute. 2000. *World Resources 2000–2001: People and Ecosystems, The Fraying Web of Life.* Washington, DC: World Resources Institute.

UN Environment Programme. 2002a. *Africa Environment Outlook.* Hertfordshire, UK: Earthprint Limited and UNEP.

———. 2002b. *Global Environment Outlook 3 (GEO-3).* London: UNEP and Earthscan.

UN Environment Programme, Collaborating Centre on Energy and Environment. 1997. "Implementation Strategy to Reduce Environmental Impact of Energy Related Activities in Zimbabwe." Working Paper No. 5. Risø National Laboratory, Denmark (January).

Veit, Peter G. 1998. *Africa's Valuable Assets: A Reader in Natural Resources.* Washington, DC: World Resources Institute.

World Bank. 2001. *Middle East and North Africa Region Environmental Strategy: Towards Sustainable Development.* Washington, DC: World Bank.

World Commission on Dams. 2000. "Large Dams and Their Alternatives in Africa and the Middle East: Experiences and Lessons Learned." Available at www.dams.org/kbase/consultations/afrme.

———. 2001. "Africa: Irrigation and Hydropower Have Been the Main Drivers for Dam Building. Dams and Water: Global Statistics." Available at www.dams.org.

Zarsky, Lyuba, ed. 2001. *Human Rights and the Environment.* London: Earthscan.

9

Air Quality
and the Atmosphere
—Mary Krane Derr

In numerous regions of Africa, human, plant, and animal communities exist in an environment largely unsullied by accumulations of airborne pollutants. Indeed, in areas that have seen little industrial development or other incursions by people, air quality remains very high. This quality extends to most rural communities as well, although indoor air pollution from cooking and heating practices has garnered increased attention in recent years. In the continent's proliferating metropolitan areas, however, addressing air pollution has emerged as a top priority, even though individual countries—and the continent as a whole—produce small amounts of air pollutants by international standards. In the Middle East, meanwhile, a similar situation is unfolding. Low levels of industrialization have historically reined in emissions of air pollutants in the region, but population growth, urbanization, and an enormous complex of oil-related industries have produced air pollution "hot spots" in large cities and other centers of oil industry activity (UN Environment Programme 2002b).

Air Quality in Africa and the Middle East
Air quality in undeveloped, modestly populated regions of Africa and the Middle East is generally good, except for localized episodes of land clearance by burning. Indeed, emissions of carbon dioxide and other chemicals from the nations that compose this region are modest when compared with those of other parts of the world. But in both Africa and the Middle East, unbridled urbanization is eroding air quality in major population centers at an alarming rate.

Most of this deterioration in air quality has taken place in the last half-century. During that time, emissions of destructive chemical compounds into the atmosphere from human activities have escalated dramatically in urban areas, causing numerous environmental and health problems. In addition to air pollution, acid precipitation, thinning of the stratospheric ozone layer, and warming of the global climate have been other major consequences of this surge in emissions. In many parts of the world, these airborne chemicals, produced primarily from the burning of oil and coal, have accumulated to the point that they pose a potential threat to human communities and natural ecosystems alike. All of these maladies are evident to one degree or another in Africa and the Middle East.

Air Pollution in African Urban Centers

Africa's rate of urbanization of 3.5 percent per year is believed to be the highest in the world, even exceeding that of Asia, another continent experiencing galloping population growth. According to the UN Centre for Human Settlements, there are currently forty cities in Africa with populations of 1 million or more, and it is expected that by 2015, this number will increase to seventy. Urbanization is proceeding at an especially brisk pace in northern Africa, central Africa, and the West Indian Ocean Islands. Western and southern Africa, by contrast, have a greater percentage of their total population in rural areas. But even here, difficult economic and social conditions in rural areas and the quest for a better way of life have spurred a general upward trend in urbanization (UN Centre for Human Settlements 1996).

Among individual countries, South Africa is the leading generator of common anthropogenic (manmade) chemical compounds entering the atmosphere. In 1998, for example, South Africa contributed 42 percent of the total regional emissions of carbon dioxide, the main greenhouse gas responsible for global climate change. Another three countries—Egypt, Nigeria, and Algeria—combined to account for another 35.5 percent of the continent's total emissions from fossil fuels (Carbon Dioxide Information Analysis Center 2001). All other countries emit only negligible amounts of emissions by global standards.

But these modest overall emission figures mask serious problems in some cities, where industrial activity, energy generation, automobile and truck use, and household-related emissions are concentrated. Emissions from factories, energy plants, and other industrial facilities are the most obvious culprits in declining air quality in Africa's cities. Billowing forth from smokestacks in thick clouds, the chemical compounds generated in these facilities sometimes include asbestos, heavy metals, and other cancer-causing agents, as well as particulate matter and sulfur dioxide, which play a part in the generation of atmospheric smog.

Africa's continued heavy reliance on coal for energy is a particularly troubling obstacle to improving air quality, since coal releases sulfur dioxide and nitrous oxides into the air when consumed. The environmental damage from Africa's coal consumption could be reduced—though not eradicated—through the introduction of new pollution-abatement technologies that have been introduced in the industrialized world. But many industries and energy plants have not made such investments. In North Africa, for instance, the current industrial capacity rests in large measure on factories that were built during the 1960s. These old facilities generate appalling amounts of air pollution, but many continue to operate unfettered by government constraints. "Protective trade regimes, foreign exchange constraints, and . . . dominance of the public sector in industry . . . provide little incentive for adopting more efficient and cleaner industrial technologies," explained the UN Environment Programme (UN Environment Programme 2002a). As a result, many cities in northern Africa that are close to refineries and oil-fired power plants using high-sulfur fuel now register sulfur dioxide levels in the air that are more than twice the level of World Health Organization safety standards (World Bank 2001d).

Another vexing problem for many African urban centers is the surge in emissions from motor vehicles, which are rapidly increasing in number across much of the continent. Emissions from automobiles, motorcycles, trucks, and buses are Africa's major source of lead contamination, and they generate large volumes of carbon dioxide and other greenhouse gas emissions linked to global climate change. These emissions are proving difficult to curb, for many African nations heavily tax fuels and new vehicle imports, and affordable spare parts are a rarity in some regions. Citizens must thus rely upon older, inadequately maintained vehicles and cheaper but more polluting fuels, especially diesel and leaded gasoline. In northern Africa, for example, the number of motor vehicles in most countries has nearly doubled over the past fifteen years, but this expanding vehicle fleet consists primarily of older models. These older cars and trucks emit twenty times more hydrocarbons and carbon monoxide and four times more nitrogen oxides than new vehicles. In addition, studies indicate that particulate emissions from poorly maintained diesel-fueled buses and trucks are five to seven times higher than those from well-maintained vehicles (Larson 1995; UN Environment Programme 2002a).

Diesel exhaust particulates are notoriously heavy and toxic. Leaded gasoline has disappeared from most of the planet, but not from Africa, which has the world's highest rate of lead poisoning. Lead and other toxic metals can cause organ damage as well as brain and nervous system impairment, particularly in the very young. In Kampala, Uganda, vehicular emissions are believed

Workers heading for a chemical fertilizer plant on the outskirts of Alexandria, Egypt. CORBIS

to account for 95 percent of lead concentrations in the air (National Environmental Management Authority 2001).

Rapid urbanization also means rapid waste accumulation. Garbage collection may not exist in cities' informal settlement areas ("shantytowns"), where the rural poor frequently migrate. Frequently, these mountains of rubbish are burned off, producing heavy albeit localized concentrations of air pollutants. Some African cities such as Addis Ababa, Ethiopia, do have municipal landfills, but these emit methane, a recognized greenhouse gas. Only hospital waste has been routinely incinerated, and medical incineration is as controversial as any other kind (Health Care Without Harm 2001). Incinerators emit dioxins, furans, mercury, lead, particulates, sulfur dioxide, carbon monoxide, and nitrogen oxides. Although some advocate incineration as an economic boon, organized opposition to the practice has coalesced in southern Africa and elsewhere (UN Environment Programme 2002a).

Domestic fuel burning, meanwhile, degrades indoor and outdoor air quality throughout Africa. Unable to purchase electricity because of impoverishment or infrastructure shortfalls, many Africans use coal, paraffin, or traditional (i.e., noncommercial) biomass fuels such as wood, charcoal, dung, and crop residues in their homes, especially in rural areas. Biomass, a labor-intensive and energetically inefficient fuel, constitutes almost 65 percent of Africa's total

Slum living conditions in Roche Santiero Market, Luanda, Angola. JEREMY HOMER/CORBIS

energy consumption, and in some locales heavy reliance has led to deforestation and land degradation. Most of this consumption is for cooking purposes, but heating also accounts for a significant amount of consumption in some areas. When burned, these domestic fuels release sulfur dioxide, carbon dioxide, nitrogen oxides, dioxins, hydrocarbons, and particulates. Women and young children are the chief victims of this unhealthy dynamic. Numerous studies indicate that exposure contributes to mortality from respiratory disease in young children and women, and is also a factor in low birth weight and other health problems (Koning 1990).

Air Quality in the Arab World

Middle Eastern patterns of urbanization and industrialization are chiefly responsible for declining air quality in the region. Middle Eastern urbanization rates presently meet or exceed the global average. In 1950, 25 percent of the Arab world lived in cities. By 2000 the figure was 50 percent, due largely to the expansion of petroleum, cement, power, and other industries (UN Development Programme 2002). Studies indicate that these metropolitan areas are bearing the brunt of the region's air pollution problems. Industrial and energy sectors remain inefficient users of fossil fuels, and their pollution loads are exacerbated by heavy industrial emissions of particulates and sulfur oxides and significant gas flaring activity in countries such as Saudi Arabia (World Bank 2001b). For example, in Al-Mashriq (the Levant), the Arabic countries bordering the East Mediterranean Sea, cement plants release large quantities of dust, soot, and sand that have been blamed for smothering nearby vegetation and afflicting workers and community residents with serious respiratory diseases. Meanwhile, the main source of air pollution in the Gulf Cooperation Council (GCC) countries—Bahrain, Kuwait, Oman, Saudi Arabia, Qatar, and United Arab Emirates—is the economically vital oil industry. "In the Mashriq countries, outdated technologies especially in power generation plants, fertilizer plants, smelters, and cement factories have caused deterioration of air quality not only in industrial sites but also in nearby settlements" (UN Environment Programme 2002b). Instituting measures to curb these glaring problems will be difficult, as weak regulatory agencies and anemic enforcement mechanisms are the rule rather than the exception across much of the region.

Reducing airborne pollutants from the Middle East's transportation sector will also be a challenge. Characteristics of the transport sector in most countries include a growing vehicle fleet, large numbers of old cars and trucks that generate high emissions of pollutants, and heavy traffic congestion in cities and high-traffic corridors. Unleaded gasoline is slowly making inroads into the region, but conversion has been slow. Finally, seasonal sand and dust storms contribute to air pollution in West Asia in general and along the northern coasts of the Arabian (Persian) Gulf in particular. In many cases these storms expose people to chemicals picked up from fertilizer or pesticide applications (ROPME 1999).

Climate and Conflict among Leading Mitigating Factors

Unique natural features of local landscape and climate can either reduce or compound the impact of air pollution from any source. In the winter especially, local atmospheric conditions cause smoke from millions of domestic fuel users in Cape Town's informal settlement areas to stagnate over the city instead of blowing out to sea. The island of Cyprus, on the other hand, retains little pollution other than ozone from vehicle exhaust and power generation. On the Arabian Peninsula, natural sandstorms combine with humanmade particulates: double jeopardy for asthma sufferers. Similarly, arid climatic conditions and desertification have increased levels of airborne particulate matter in many African landscapes.

Warfare and political unrest have also contributed to degradation of air quality and other environmental elements of the Middle East in recent years. During the 1991 Persian Gulf War, for example, retreating Iraqi troops torched Kuwait's oil fields. The fires burned and smoked for seven months, consuming more than 1 billion barrels of oil. "The deposition of oil, soot, sulfur, and acid rain on croplands up to 1,200 miles in all directions from the oil fires turned fields untillable and led to food shortages that continue to this day," wrote Jonathan Lash, president of the World Resource Institute. "The fires released nearly half a billion tons of carbon dioxide . . . emissions greater than all but the eight largest polluting countries for 1991 that will remain in the atmosphere for more than a century. The oil that did not burn in the fires traveled on the wind in the form of nearly invisible droplets resulting in an oil mist or fog that poisoned trees and grazing sheep, contaminated fresh water supplies, and found refuge in the lungs of people and animals throughout the gulf" (Lash 2002).

Addressing Regional Air Quality Issues

Strategies for improving air quality at both the urban and rural levels in the Middle East and Africa hinge on imposing new restrictions on emissions in all sectors (industrial, energy, transport, and household); improved enforcement of existing regulations; improved research and monitoring of air quality; greater integration of environmental considerations in other aspects of government, municipal, and industry planning; and improved international cooperation in addressing transboundary air pollution and other issues.

In the Middle East, movement is already being seen—especially in the GCC countries—toward adoption of cleaner production approaches in industry, especially in the large oil, petrochemical, fertilizer, and metal industries (UN Environment Programme 2002b). Most Arab governments now have ministries of environment, although their mandate remains primarily one of public education, a function they share with various environmental nongovernmental organizations (NGOs) with a presence in the region. Indeed, international organizations are playing an important role in addressing air quality issues. For example, urban planning to reduce traffic and pollution is a chief objective of the World Bank's Sub-Saharan Africa Transport Policy Programme. It figures as well in the reconstruction of the Central Business District (CBD) of Beirut, Lebanon, after two decades of civil war. The CBD redesign places homes close to workplaces and supports pedestrian-friendly public spaces and citywide mass transit (UN Environment Programme 2002a).

In Africa, multistate agreements focused on air pollution issues are proliferating, with transport emissions the focus of particularly ambitious activity. In 2001, for example, twenty-five sub-Saharan African nations passed the Declaration of Dakar, an action plan to phase out leaded gasoline by 2005. Many African and Middle Eastern nations have also committed to the UN

Motor Vehicle Emissions Agreement, which aspires to reduce pollution levels, improve vehicle safety via the introduction of globally uniform standards, and promote fuel efficiency. At the national level, Lebanon and the Gulf petroleum states are transitioning to unleaded gasoline and catalytic converters (devices to remove carbon monoxide and hydrocarbons from vehicle exhaust). Dubai's government vehicle fleet is now hydrogen-powered, and Egypt ended all sales of leaded gasoline in 2002. Israel and Senegal have restricted the import of older, more polluting vehicles, and in North Africa, relaxed trade policies have encouraged the replacement of older vehicles with affordable new ones (World Bank 2001a; UN Environment Programme 2002a; Sperling and Salon 2002).

Ozone Depletion: An Effective International Response to a Global Issue

Advocates of strong action to curb emissions responsible for global climate change contend that the world's recent efforts in the realm of stratospheric ozone preservation constitute an encouraging example of humankind's capacity to redress environmentally destructive behavior. Indeed, the international response to protect the stratospheric ozone layer from ozone-depleting chemicals over the last two decades has been both decisive and effective (UN Environment Programme 2000; World Meteorological Organization and UN Environment Programme 1998).

Efforts to preserve the ozone layer from thinning and loss are critically important for several reasons. Ozone in the stratosphere (between six and thirty miles above the surface) protects earth life from the full force of the sun's ultraviolet radiation, which is a cancer-causing agent. In addition, other harmful effects associated with increased exposure to ultraviolet radiation include damage to valuable food crops and other vegetation and declines in plankton, a vital link in the world's marine food chain. These losses can in turn lead to escalating levels of carbon dioxide—the primary cause of global warming— and prompt dramatic changes in the character of regional ecosystems.

In the 1980s, evidence of ozone loss and associated environmental problems became too overwhelming for the world to ignore. In response, many of the world's industrialized nations united to pass the 1987 Montreal Protocol on Substances that Deplete the Ozone Layer, the first international treaty explicitly geared toward protection of the global atmosphere. Now ratified by more than 180 countries, the 1987 Montreal Protocol ordered a global phaseout in production and consumption of ozone-depleting substances (ODS).

Many African countries, including Egypt, Ghana, Kenya, Senegal, and Togo, were among the protocol's first signatories, and over 80 percent (forty-five of fifty-three nations) have now signed on to the treaty. In addition, African countries have, proportionately, one of the best records in ratifying subsequent amendments to the Montreal Protocol, and 70 percent of African parties to the protocol have fully complied with reporting requirements. Most important, Africa has made dramatic reductions in its consumption of ozone-depleting substances throughout the industrial and commercial sectors with the help of various international institutions and nongovernmental groups (UN Environment Programme 2000).

Africa's decisive response to the crisis has earned special praise from the world community. "The ozone layer crisis is one of many global environmental problems primarily caused by developed countries," observed the UN Environment Programme. "But knowing that a global environmental threat can only be solved through the commitment and participation of all regions and nations, Africa has contributed beyond all expectation. . . . The contribution is all the more commendable, given the fragile nature of Africa's economies and political stability. . . . [Yet] Africa was still able to muster the political will and leverage the financial resources needed for progress" (ibid.).

Since Montreal, global CFC production and consumption have fallen more than 85 percent—95 percent in industrialized countries—and while stratospheric concentrations may be about to peak because of emissions generated in earlier years, tropospheric ODS levels are waning (ibid.). Still, scientists contend that the ozone layer—and the planet it protects—will remain vulnerable unless developing nations fully comply with the protocol and illegal trade in ODS is halted. In Africa, for example, industrial pockets saddled with old and polluting machinery continue to spew high volumes of ozone-depleting substances into the air. Moreover, efforts to ban the use of ozone-depleting substances such as methyl bromide, which is widely used as a fumigant in agriculture, have floundered in some countries because of a paucity of cheaper or more effective alternatives and governmental apathy. In addition, the international community needs to address environmental problems associated with some ODS alternatives, such as hydrochlorofluorocarbons (HCFCs) and hydrofluorocarbons (HFCs), two CFC substitutes that are now known to contribute to global warming (Anderson and Sarma 2003).

Global Climate Change and Its Impact on Africa and the Middle East

The Dramatic Natural "Baseline" and Its Dramatic Exaggeration

Throughout recorded history, Africa has been known for its wide-ranging climates and extreme weather events such as sandstorms, droughts, floods, and cyclones. Over the centuries, the arid and semiarid countries of Africa's north, south, and Horn have experienced repeated drought episodes as well as major floods. A key component in the continent's climate is the El Niño Southern Oscillation (ENSO), which helps shape the weather in the Western Indian Ocean Island nations and across most of east, west, central, and southern Africa. ENSO is a vast, complex cycle of fluctuations in ocean temperatures, rainfall, atmospheric circulation, and air pressure across the tropical Pacific. It affects meteorological events as far as half a world away.

The peoples of the Middle East are also accustomed to challenging climatic conditions. Much of this region, including most of the Arabian Peninsula and Israel, typically sees exceedingly light precipitation (less than ten inches of annual rainfall), extremely hot daytime temperatures, and cold nights that can go below freezing. Except for nighttime and sandstorms, little interrupts the intense sunlight. The Mediterranean basin, meanwhile, features rainy and mild winters and hot and dry summers.

During the 1990s, the earth experienced three times as many weather-related disasters as it did during the 1960s, with ten times the economic losses (Intergovernmental Panel on Climate Change 2001a). Some have attributed this upsurge of disasters—at least in part—to population pressures that cause mismanagement of land and water resources and make them more vulnerable to severe weather events (Mahoney 2000). But most members of the scientific community also cite changing weather patterns stemming from a buildup of anthropogenic greenhouses—commonly known as "global warming"—as a major factor in the intensification of some weather characteristics. For example, average precipitation rates have clearly declined in Africa since 1968, despite an increase in flash flood events, and drought conditions have appeared more frequently in perennially arid nations such as Botswana, Burkina Faso, Chad, Ethiopia, Kenya, Mauritania, and Mozambique (UN Environment Programme 2002a).

Many African communities and economies have suffered cruelly from these blows. "Malnutrition and famine have resulted from both droughts and floods in Africa and associated food imports and dependency on food aid have contributed to limited economic growth of the countries affected. . . . Over the past 30 years, millions of Africans have sought refuge from natural disasters, often settling in fragile ecosystems and/or experiencing social tensions with neighboring communities. Ecological impacts of drought and flooding include land degradation and desertification, loss of natural habitat or changes in distribution of biodiversity, increased soil erosion and silting of rivers, dams, and coastal ecosystems" (ibid.). These tribulations, combined with the HIV/AIDS pandemic, have taken a grave toll upon health, educational systems, and economic productivity in many parts of the continent

(Save the Children and Oxfam International 2001; International Federation of Red Cross and Red Crescent Societies 2002).

Although it has not experienced the same barrage of catastrophes, the Middle East is also struggling to address a host of environmental problems—including land degradation, desertification, and freshwater scarcity—that are possibly being exacerbated by recent climatic changes (Pe'er and Safriel 2000; Tsiourtis 2002).

Many of these significant weather phenomena that have rolled across Africa and the Middle East in recent years have been directly linked to ENSO. But most scientists now believe that global climate change associated with human activity is altering and exaggerating ENSO patterns, triggering a rise in adverse weather events (Intergovernmental Panel on Climate Change 2001a). For example, the 1997–1998 ENSO triggered very high sea surface temperatures in the southwestern Indian Ocean, causing high rainfall, cyclones, flooding, and landslides across most of Eastern Africa and drought conditions in southwestern Africa. The higher sea temperatures also caused extensive bleaching of corals on the eastern African coast and in the West Indian Ocean Islands (Wilkinson 2000). Many climate experts fear that these events could be a precursor of weather changes to come.

The Nature of Humanmade Climate Change

The main generators of "greenhouse gases"—carbon dioxide, nitrous oxide, chlorofluorocarbons (CFCs)—and other chemicals that trap the sun's heat in the atmosphere are various types of anthropogenic (human) activity, specifically the burning of oil, gas, and coal to operate cars, trucks, airplanes, factories, and power plants. The consumption of fossil fuel generates huge quantities of carbon dioxide, the main greenhouse gas. Lesser sources of greenhouse gases include methane emissions from livestock and landfills, nitrous oxides from agricultural fields, emissions of fluorinated gases from industry, emissions of carbon dioxide from volcanic activity, and releases of carbon dioxide from the clearance of carbon-storing forests.

According to the 2,500-member Intergovernmental Panel on Climate Change (IPCC), a group operating under the joint sponsorship of the United Nations and the World Meteorological Organization, evidence of global warming attributable to human activities is accumulating around the planet, with rapid melting of glaciers and polar ice caps and record-breaking temperatures the most noteworthy manifestations. According to the IPCC, the world's most authoritative source on global warming, nine of the world's ten hottest years in recorded history occurred between 1990 and 2000 (Intergovernmental Panel on Climate Change 2001b).

Most scientists believe that the earth's accelerating retention of greenhouse gases in the atmosphere will wreak major changes on the planet and its natural ecosystems. For example, the IPCC has forecast that the planet will warm by a stunning 1.4 to 5.8 degrees Celsius (2.5 to 10 degrees Fahrenheit) over the course of the twenty-first century without major reductions in greenhouse gas emissions. Some of these sweeping changes could be beneficial to certain regions—by transforming agriculturally marginal land into productive farmland, for example—but the tidings for many peoples, animals, and plants are grim. Probable repercussions of global warming include increasingly severe and numerous storms, altered rain and snowfall patterns that will bring greater incidence of flooding and drought, inundation of islands and coastal areas from rising sea levels precipitated by melting glaciers and polar ice caps, expansion of malaria and other tropical diseases into previously temperate zones, and possible mass extinctions of species of mammals, birds, reptiles, amphibians, fish, and plants (Intergovernmental Panel on Climate Change 2001a).

Scientists believe that the severity of many of these changes could be reduced if nations emitting high levels of greenhouse gases took prompt action now. Indeed, the decisive international response to the stratospheric ozone loss issue has been cited as a model for crafting future climate policy (Downie 1995). Thus far, the main international responses to this brewing crisis have been the 1992 UN Framework Convention on Climate Change (UNFCCC) and the 1997 Kyoto Protocol. The latter UN-brokered agreement calls on developed nations to reduce their emissions of greenhouse gases to at least 5 percent below 1990 emissions levels by 2012. The protocol enters into force when it has been ratified by at least fifty-five parties to the convention, including developed countries accounting for at least 55 percent of total carbon dioxide emissions in 1990. But the future of the Kyoto Protocol, which even supporters acknowledge is only a first step in addressing global climate change, is uncertain. The United States, which ranks as the world's leading producer of greenhouse gases, has decided not to ratify the treaty, citing the protocol's exclusion of developing countries and the potential economic cost of meeting its obligations.

Greenhouse Gas Emissions in Africa and the Middle East

Before the Industrial Revolution roughly 250 years ago, earth's total output of greenhouse gases (GHGs)—mainly from respiring plants and rotting organic matter—was balanced out by the absorbing qualities of "carbon sinks" such as forests and oceans. Since then, human activities—principally fossil fuel combustion and cement manufacture—have raised atmospheric carbon dioxide levels by almost 30 percent, disturbing this ancient equilibrium. Moreover, fully 50 percent of all industrial carbon dioxide emissions produced in the entire history of the planet have been generated and released since the mid-1970s (Intergovernmental Panel on Climate Change 2001b).

Africa and the Middle East account for a relatively modest portion of the world's greenhouse gas emissions when compared with industrial giants such as the United States and the European Community. The Middle East, however, is the primary *source* for much of the oil that is consumed around the world, and many experts believe that West Africa may be the world's next major frontier of oil exploration and development.

Africa contains about 13.5 percent of the world population, but it contributes less than 3.5 percent of global emissions of carbon dioxide, the main greenhouse gas. Africa's emissions come chiefly from South Africa (which alone accounts for 42 percent of the continent's emissions), Algeria, Egypt, and Nigeria (Carbon Dioxide Information Analysis Center 2001). The rest of Africa's emissions are comparatively small. For example, the fourteen nations that compose the Southern African Development Community (SADC) account for about 2 percent of global GHGs, about half as much as the United Kingdom alone (World Wide Fund for Nature). But emissions in southern Africa and other parts of the continent are projected to rise significantly in the coming decades if present patterns of development hold true. Zimbabwe, for example, could see a threefold increase in emissions over the next fifty years (Southern Centre 1996).

The Middle East, which has less than 2 percent of the world population, is also a minor producer of carbon dioxide and other greenhouse gases when viewed in an international context. Total emissions of carbon dioxide in the region in 1998, for example, were 197 million tons; Europe, by contrast, produced 1.677 billion tons and Asia and the Pacific generated 2.167 billion tons. But on a per capita basis, the region's emissions of carbon dioxide are significant. Per capita carbon dioxide emissions in the Arab world rose from 4.7 tons per year in 1972 to 7.4 tons per year in 1998, echoing regional trends in industrialization, development, and population growth. Since World War II, population growth, urbanization, and expansion of fossil fuel–burning power stations, oil-related industry, cement manufacture, and waste disposal have dramatically increased energy consumption and atmospheric pollution levels (Carbon Dioxide Information Analysis Center 2001; UN Environment Programme 2002a; National Energy Foundation 2002).

Global Warming's Potential Impact

If global warming continues unchecked, the IPCC has quite sobering predictions for Africa and the Middle East. Africa's rainfall and temperature cycles will grow more unpredictable and variable in conjunction with ENSO events. Vegetation zones will shift. Evaporation will increase, as will vulnerability to droughts, floods, landslides, and cyclones. Indeed, the IPCC has cited Africa as the continent most vulnerable to global climate change, as endemic poverty and dysfunctional governments will make it even more difficult for nations to deal with predicted decreases in freshwater and food security (Intergovernmental Panel on Climate Change 2000).

Regions of Africa already receiving extremely limited rainfall—the Horn of Africa, the Sahel, and southern Africa—will suffer the most, according to climate change models. Because of higher rainfall, floods will occur more frequently in the equatorial belt, particularly if its rain forests—vital carbon sinks for the whole planet—continue to suffer further diminishment at the hands of loggers and agricultural interests. Tropical deserts may grow hotter. Temperature rise and changes in rainfall and evaporation rates will further stress Mediterranean zones. Continued overgrazing, unsustainable farming, and deforestation will amplify the impact of all of these changes.

Food insecurity will become more pronounced in places like the Sahel and the Horn of Africa. Land suitable for cultivation will decrease, at the same time that population increases and crop yields fall. In high-altitude regions, loss of snow and ice may eventually cut crop yields in fields that depend on runoff by as much as 20 percent, while irrigation-dependent farming, hydroelectric power generation, and tourism will also be affected. For example, the snow cap of Tanzania's Mount Kilimanjaro, the country's top tourist attraction, is melting at such a swift rate that it may be gone altogether by 2020 (Thompson 2002). Few predictions have been made for climate change effects upon Middle East economies. However, the present difficulties of overburdened, degraded, desertifying land are generally expected to intensify under developing climate regimes (Intergovernmental Panel on Climate Change 2001a).

In both Africa and the Middle East, water shortages may escalate political conflicts. Important African freshwater lakes and river systems such as the Nile and Lake Victoria will shrink, and inland fisheries, a major protein source, will diminish. Changes in rainfall could also have serious consequences for those parts of Africa that depend heavily on hydroelectricity, such as South Africa and Zimbabwe. Indeed, hydroelectric power currently generates more than 80 percent of the total power in eighteen African countries (World Commission on Dams 2001). Crop failures and drought conditions may also prompt desperate villagers to flee their parched lands for cities that are already groaning under the weight of proliferating slums, traffic gridlock, and a multitude of other environmental problems. Higher incidence of infectious diseases (such as malaria and yellow fever) and heat-related illnesses and deaths have also been forecast for Africa and the Middle East (Intergovernmental Panel on Climate Change 2001a).

Inundation of coastal shores and islands—including significant population centers—is another distinct possibility. Indeed, coastal settlements in the Gulf of Guinea, Senegal, Gambia, Egypt, and along much of the southeastern African coast would likely see varying levels of coastal erosion and outright submersion of land as a result of projected increases in sea level. For example, more than 20 percent of the people in the West Indian Ocean Islands would be at risk from a sea rise of 1 meter, and 70 percent of the Seychelles Islands land mass would disappear under the waves under such a scenario. Temperature rise could also diminish marine fisheries by damaging the health of vital reef and mangrove habitats. Finally, scientists note that even if anthropogenic greenhouse gas emissions were stabilized immediately, sea levels are likely to creep up for years to come as gases already emitted work their way through the atmosphere (ibid.).

Biodiversity and Ecosystem Integrity at Risk

Changes to rainfall and temperature patterns would also prompt extensive changes to natural ecosystems and regional biodiversity. Under many climate change models, concentrations of unique species, such as those in South Africa's Cape Floral Kingdom, will shift radically or disappear as they fail to adapt or migrate to more hospitable areas (Levine 2002). According to one study by the World Wide Fund for Nature, a mere 5 percent decrease in rainfall in southern Africa would adversely affect grazing species such as hartebeest, wildebeest, and zebra, which would in turn create a cascading impact throughout regional food chains. Reductions in tall-tree open savanna grasslands could also damage species from the sable antelope to the grass owl, while species that roam the savanna highlands, such as the gray rhebok and mountain reedbuck, would also be placed at risk. The potential for habitat changes of this magnitude is of grave concern to conservationists. According to WWF analyses, between 15 and 20 percent of the large nature reserves in southern Africa—including species-rich areas such as Kruger National Park, the Zambezi basin, and the Okavango Delta—would experience a change in biome (habitat type) under any of three likely climate change scenarios (World Wide Fund for Nature).

Other internationally significant centers of biodiversity would also be jeopardized under current climate change forecasts. For example, Lake Malawi National Park's namesake lake, which is shared by Malawi, Tanzania, and Mozambique, supports one of the world's most productive fisheries. As many as 1,000 fish species prowl its waters, many of them endemic in character. But changing climatic conditions will likely wreak major alterations to the lake's ecosystem. "Global warming is likely to increase the temperature and density difference between the surface and deep waters, causing the lake to become more vertically-stratified," observed the WWF. "This will reduce the mixing of the nutrient-rich deep water and the nutrient-poor surface waters, which is fundamental to maintaining the fisheries. A decline in the fish population would be disastrous not only for the conservation of the unique cichlid (a family of perch-like fishes) fauna, but also for the local people who depend on the fishery for food, not to mention fish eagles, white breasted and reed cormorants, pied kingfishers, spotted-necked and clawless otters, monitor lizards and other species which prey upon the fish" (World Wide Fund for Nature).

Prevention and Adaptation: The Order of the Day

Throughout the region, governments, international agencies, NGOs, and the public are engaged in multiple adaptation and prevention efforts concerning anthropogenic climatic change. This will pose a formidable challenge, however, as nations in both Africa and the Middle East confront a situation in which they will be seeking to mitigate the impacts of climate change by reducing their dependence on fossil fuels while simultaneously working to improve living standards and realize other economic development goals (UN Environment Programme 2002a; Rowlands 1998).

Nonetheless, most African and Middle Eastern nations are parties to the UN Framework Convention on Climate Change (UNFCCC). In addition, all African countries, with the exception of Angola, Liberia, and Somalia, have ratified its legal instrument, the Kyoto Protocol, which in its own words aspires to cap atmospheric GHGs at levels "that would prevent dangerous anthropogenic interference with the climate system, within a timeframe sufficient to allow ecosystems to adapt naturally to climate change, to ensure that food production is not threatened and to enable economic development to proceed in a sustainable manner." Developing countries have negotiated funding and a delayed GHG reduction timetable for themselves, but the Kyoto mechanisms and international institutions to be created to realize those mechanisms do provide for the active participation of developing countries in Africa and elsewhere—through, for example, increased adoption of energy-efficient and renewable energy technologies (UN Environment Programme 2002b; UN Framework Convention on Climate Change).

If the Kyoto Protocol eventually receives enough support to come into force, it would be enormously beneficial for the African continent. Passage would signal international determination to preserve ecosystems and flora and fauna in Africa and around the world, as well as recognition of the unfathomable threat that climate change poses to the continent's already fragile socioeconomic fabric. Moreover, the ascension of the protocol into force would deliver funding to African countries through such programs as the Special Climate Change Fund and the Fund for Least Developed Countries. As the UN Environment Programme explains: "[U]nder the Protocol's mechanisms, developed countries will be able to offset some of their emissions by paying for carbon-saving projects such as tree planting and forest conservation schemes in developing countries. Funds will also be available to help developing countries to convert to cleaner technologies such as solar and wind power or fuel-cell-operated vehicles, currently too expensive for many African nations" (UN Environment Programme 2002a).

Besides Kyoto, however, African countries can take a number of steps to reduce GHG emissions and prepare for climatic changes that are on the horizon. For example, recent measures to improve air quality in the region will also have the beneficial effect of addressing the looming global warming issue, as many air pollutants are also greenhouse gases. In addition, programs designed to address a host of environmental issues, from desertification to urban sprawl, have been expanded to include climate change considerations, with both prevention and adaptation components. Other areas that African countries need to prepare for include changes in water resource distribution, rangeland and protected area management, and disease distribution. To that end, experts have urged policy-makers to tap into the knowledge of indigenous, rural communities. "For centuries, rural communities have learned to survive drought and harsh weather in Africa, but many of the water and soil management techniques, resistant crop varieties, and food production methods are known only locally, or to certain ethnic groups. These tried and true adaptive strategies need to be extended out of their areas of origin to achieve wider application" (World Wide Fund for Nature). Exploration of potential new crop varieties is also underway, including genetically modified seed strains that can prosper in arid regions. This potential avenue of adaptation is not without controversy, however, as scientists, environmentalists, and policy-makers around the world remain engaged in a fierce debate about the health and environmental benefits and drawbacks of GM agriculture systems.

Research, monitoring, and public education efforts are underway as well. For example, the AGRHYMET Regional Centre, a joint effort of nine Sahel countries, educates farmers about food security, livestock management, and plant protection, and the West Indian Ocean Islands take part in the Global Ocean Observing System, which generates research data used in climate change modeling.

Other strategies involve preserving and restoring natural habitats, biodiversity, and carbon sinks, along with the rural livelihoods that depend upon them. For example, one mandate of the Central African Regional Programme for the Environment is to address forest and biodiversity conservation in the Congo Basin, which ranks second only to South America's Amazon region as a planetary carbon sink.

Factories in North Africa have already begun to reduce carbon dioxide output, through emissions controls or a shift to natural gas power, and Madagascar, Morocco, Algeria, and Egypt are all investing in wind, solar, and waste-to-energy initiatives (UN Environment Programme 2002a).

In the Middle East, governments and UN agencies have sought business and public support for greenhouse gas cuts through "cleaner production" initiatives. These efforts include reductions in the consumption of natural resources; greater restrictions on the use of highly toxic or environmentally harmful substances; improved product and manufacturing design; and increased emphasis on recycling. Already, some Gulf state industries have made meaningful cuts in their GHG emissions by improved observation of existing air quality laws, imposition of new air quality regulations, and shifts to cleaner modes of energy and industrial production.

The Middle East's leading petroleum states are also exploring alternatives to their current oil-based socioeconomic structures. The Arabian Peninsula has extensive, mostly untapped reserves of cleaner-burning natural gas and a potential abundance of renewable sun, wind, and hydrogen resources at their disposal. Indeed, the United Arab Emirates is already generating hydrogen power from seawater and arraying solar panels in the desert (Ministry of Information and Culture 2003). Rooftop solar water heaters are already quite common in Israel, and more than 90 percent of Cypriot households have used them since the early 1960s (Small Island Developing States Network, n.d.). In addition, photovoltaic cell systems are being promoted in Egypt, Israel, Palestine, and Jordan for rural villages and other "off-grid" communities—those not linked to existing utility sources.

Climate change adaptation and prevention in Africa and the Middle East depends upon responses from all sectors of society, and upon the willingness of more developed regions of the world to curb their GHG emissions and offer funds and technical assistance to developing nations. But successful mitigation and adjustment will also require fundamental changes in the way that some states operate and function. Indeed, individual countries' capacity to weather the diverse and complex changes associated with global warming will hinge on strengthening national environmental laws; expansion of the social "safety net" to account for citizens suffering economic or physical dislocation;

increased recognition of the assistance that NGOs can provide in enforcement and monitoring of environmental regulations or aid programs; greater harmonization of national environmental assessment regulations with international standards; and effective promotion of—and institutional support for—clean technologies and environmentally friendly business practices (World Bank 2001b).

Sources:

African Energy Policy Research Network/Foundation for Woodstoves Dissemination. AFREPREN/FWD website. Last updated December 12, 2002. Available at http://www.afrepren.org/ (accessed January 2003).

Air Pollution Information Network Africa. APINA website. Last updated December 18, 2002. Available at http://www1.york.ac.uk/inst/sei/rapidc2/ember 2002 (accessed January 2003).

Anderson, Stephan, and K. Madhava Sarma. 2003. *Protecting the Ozone Layer*. London: Earthscan.

Carbon Dioxide Information Analysis Center. 2001. *Global, Regional, and National Fossil Fuel CO_2 Emissions*. Oak Ridge, TN: Oak Ridge National Library, DOE.

Downie, David. 1993. "Comparative Public Policy of Ozone Layer Protection." *Political Science* 45 (December).

———. 1995. "Road Map or False Trail: Evaluating the Precedence of the Ozone Regime as Model and Strategy for Global Climate Change." *International Environmental Affairs* 7 (fall).

———. 1999. "The Power to Destroy: Understanding Stratospheric Ozone Politics as a Common Pool Resource Problem." In J. Samuel Barkin and George Shambaugh, eds., *Anarchy and the Environment: The International Relations of Common Pool Resources*. Albany: State University of New York Press.

Gabbay, Shoshana. 1999. "Climate Change: Towards Reduction of Greenhouse Gases." *Israel Environment Bulletin* 22 (summer).

Gavin, Angus. "Leaded Gasoline Phase-Out in Africa." Environmental Global Lead Network. Available at http://www.globalleadnet.org/policy_leg/policy/africa.cfm (accessed January 2003).

Health Care without Harm. 2001. *Non-Incineration Medical Waste Technologies*. Washington, DC: Health Care without Harm.

Intergovernmental Panel on Climate Change. 2000. *The Regional Impacts of Climate Change: An Assessment of Vulnerability*. Geneva: IPCC.

———. 2001a. *Climate Change 2001: Mitigation, Impacts, Adaptation, and Vulnerability: Summaries for Policymakers*. Geneva: IPCC.

———. 2001b. *Climate Change 2001: The Scientific Basis*. Geneva: IPCC.

International Federation of Red Cross and Red Crescent Societies. 2002. *World Disasters Report 2002: Focus on Reducing Risk*. Available at http://www.ifrc.org/publicat/wdr2002/ (accessed January 2003).

Israel, State of. 1999. "Israel's Air Resources Management Program." Available at http://www.israel-mfa.gov.il/mfa/go.asp?MFAH00ib0.

Koning, Henk W. de. 1990. "Air Pollution in Africa." *World Health* (January–February).

Larson, B. 1995. *Natural Resource Extraction, Pollution, Intensive Spending and Inequities in the Middle East and North Africa.* Washington, DC: World Bank.

Lash, Jonathan. 2002. "The Environment: Another Casualty of War?" World Resources Institute. December. Available at http://jlash.wri.org/letters.cfm?ContentID=564 (accessed January 2003).

Levine, Ketzel. 2002. "A Journey to the Cape Floral Kingdom National Park." National Public Radio broadcast transcript, October 23. Available at http://www.npr.org/programs/talkingplants/features/2002/southafrica (accessed January 2003).

Mahoney, Alex. 2000. "Environmental Degradation and Disasters." U.S. Agency for International Development. Available at http://www.usaid.gov/hum_response/ofda/00annual/environmental.html (accessed January 2003).

Ministry of Information and Culture, United Arab Emirates. 2003. "Environment and Energy 2003: A Major Event in Arab Region." January 20. Available at http://www.uaeinteract.com.

National Energy Foundation. 2002. "Carbon Dioxide Emissions by Countries." January 29. Available at http://www.natenergy.org.uk/co2mment.htm.

National Environmental Management Authority. 2001. *State of the Environment Uganda 2000.* Kampala: NEMA.

Network for the Detection of Stratospheric Change (NDSC). 2003. "Connections between Climate Change and Ozone Depletion." January. Available at http://www.ndsc.ncep.noaa.gov/climchng.html (accessed January 2003).

Oxfam International. 2002. "Southern Africa Food Crisis." December 19. Available at http://www.oxfam.org/eng/campaigns_emer_safrica.htm (accessed January 2003).

Pe'er, Guy, and Uriel N. Safriel. 2000. "Effect of Future Climate Change in Israel." Israel Space Agency-Middle East Interactive Data Archive. October. Available from http://www.nasa.proj.ac.il/Israel (accessed January 2003).

Population Reference Bureau. 2002. *World Population Data Sheet 2002.* Available at http://www.prb.org/pdf/worldpopulationDS02_eng.pdf (accessed January 2003).

ROPME-Regional Organization for the Protection of the Marine Environment. 1999. *Regional Report of the State of Environment.* Kuwait City: ROPME.

Rowlands, I. H., ed. 1998. *Climate Change Cooperation in Southern Africa.* London: Earthscan.

Salloum, Habeeb. 2001. "The Flowering of Agriculture and Forestry in the United Arab Emirates." Al-Hewar Center for Arab Culture and Dialogue. Available at http://www.alhewar.com/habeeb_salloum_uae_flowering_agriculture.htm (accessed January 2003).

Save the Children and Oxfam International. 2001. "HIV/AIDS and Food Security in Southern Africa." December 1. Available at http://www.oxfam.org/eng/pdfs/pp021127_aids_safrica.pdf (accessed January 2003).

Small Island Developing States Network. "Extensive Use of Solar Water Heaters in Cyprus." n.d. Available at http://www.sidsnet.org/successstories/32.html (accessed January 2003).

Sokona, Youba. 2002. "Energy in Sub-Saharan Africa." Available at http://www.heliointernational.org/Helio/anglais/reports/africa.html (accessed January 2003).

Southern Africa Development Community, World Conservation Union-IUCN, Southern African Research and Documentation Centre, and Zambezi River Authority. 2000. *State of the Environment in the Zambezi Basin 2000.* Harare, Zimbabwe: SADC, IUCN, SARDC, ZRA.

Southern Centre. 1996. *Climate Change Mitigation in Southern Africa.* Harare, Zimbabwe: Southern Centre.

Sperling, Daniel, and Deborah Salon. 2002. *Transportation in Developing Countries: An Overview of Greenhouse Gas Reduction Strategies.* Prepared for the Pew Center on Global Climate Change. Washington, DC: May.

Stockholm Environment Institute. 1999. *Atmospheric Environment Issues in Developing Countries.* Stockholm: SEI.

Tawfiq, Nizar. "Red Sea and Gulf of Aden: Protecting Our Shared Treasures." UN Environment Programme. Available at http://www.unep.ch/seas/main/persga/redcap.html (accessed January 2003).

Temba, Peter. 2002. "Tanzania Acts to Save Kilimanjaro." Africaonline.com. November 11. Available at http://www.africaonline.com/site/Articles/1,3,49376.jsp (accessed January 2003).

Thompson, Lonnie G. et al. 2002. "Kilimanjaro Ice Core Records." *Science* 298 (October 18).

Tsiourtis, Nicos X. 2002. "Cyprus-Water Resources: Planning and Climate Change Adaptation." IUCN Mediterranean Region, November. Available at http://www.iucn.org/places/medoffice/Documentos/Cyprus.pdf (accessed January 2003).

UN Centre for Human Settlements. 1996. *An Urbanizing World: Global Report on Human Settlements 2001.* Nairobi: UNCHS.

UN Development Programme. 2000. *World Energy Assessment: Energy and the Challenge of Sustainability.* New York: UNDP.

UN Development Programme, Regional Bureau for Arab States. 2002. *Arab Human Development Report.* Available at http://www.undp.org/rbas/ahdr/english.html (accessed January 2003).

UN Development Programme, UN Environment Programme, World Bank, and World Resources Institute. 2000. *World Resources 2000–2001: People and Ecosystems, The Fraying Web of Life.* Washington, DC: World Resources Institute.

UN Environment Programme. 1998. *Production and Consumption of Ozone Depleting Substances 1986–1996.* Nairobi, Kenya: UNEP Ozone Secretariat.

———. 2000. *Action on Ozone, 2000.* Nairobi: UNEP.

———. 2002a. *Africa Environment Outlook.* Hertfordshire, UK: Earthprint Limited and UNEP.

———. 2002b. *Global Environment Outlook 3 (GEO–3).* London: UNEP and Earthscan.

UN Environment Programme, Regional Office for Africa. 2001. *Patterns of Achievement: Africa and the Montreal Protocol.* Nairobi: UNEP-ROA.

———. 2002. *Assessment Report for the WSSD Preparatory Conference.* Nairobi: UNEP-ROA.

UN Framework Convention on Climate Change. *Greenhouse Gas Inventory Database.* Available at http://ghg.unfccc.int/ (accessed January 2003).

Wilkinson, Clive, ed. 2000. *Status of Coral Reefs of the World: 2000.* Townsville, Australia: Australian Institute of Marine Science and Global Coral Reef Monitoring Network.

World Bank. 2001a. "Declaration of Dakar, June 26–28. Regional Conference on the Phasing Out of Leaded Gasoline in Sub-Saharan Africa." Available at http://www.worldbank.org/cleanair/caiafrica/africaenglish/learningactivities/dakar/conclusions/declaration/english.pdf (accessed January 2003).

———. 2001b. *Middle East and North Africa Region Environmental Strategy: Towards Sustainable Development.* Washington, DC: World Bank.

———. 2001c. *Middle East and North Africa Region Environmental Strategy Update.* Washington, DC: World Bank.

———. 2001d. *World Development Indicators 2001.* Washington, DC: World Bank.

World Commission on Dams. 2001. "Africa: Irrigation and Hydropower Have Been the Main Drivers for Dam Building. Dams and Water: Global Statistics." Available at www.dams.org.

World Meteorological Organization and UN Environment Programme. 1998. *WMO/ UNEP Scientific Assessment of Ozone Depletion.* Geneva: WMO/UNEP.

World Wide Fund for Nature. *Climate Change and Southern Africa.* Available at http://www.panda.org/resources/publications/climate/Africa_Issue/africa.htm (accessed January 2003).

———. *Protected Areas at Risk.* Available at http://www.panda.org/resources/publications/climate/parks/ (accessed January 2003).

Environmental
Activism

—Phia Steyn

Despite the fact that the future and survival of the people of Africa and the Middle East are inextricably linked to the natural environment in which they live, modern environmental activism is a fairly recent phenomenon in the two regions. A strong environmental movement did not emerge in these regions until the 1980s, when dismal economic, social, and environmental conditions and the inability of national governments to meet their obligations toward society spurred citizens into collective action to achieve environmental and other goals.

Like its counterparts in many other developing regions, environmental activism in Africa and the Middle East rarely confines itself only to the achievement of environmental goals. On the contrary, environmental struggles in the two regions more often than not form part of a broader struggle for better political, economic, social, developmental, *and* environmental rights, the reason being that environmental marginalization in Africa and the Middle East is closely linked to the political, economic, social, and developmental marginalization of both rural and urban communities in the two regions. As a result, environmental activism in the two regions has a strong developmental approach that aims at improving conditions at a local level to the advantage of local communities.

The Roots of Modern Environmental
Activism in Africa and the Middle East
The roots of modern environmental activism in Africa date back to the conservation movement that emerged during the colonial era of the late nineteenth century and first half of the twentieth century. The African environment and the wildlife contained therein captured the popular imagination of the

229

European colonizers. Africa became the proverbial Eden that had ceased to exist in Europe as a result of industrialization, the overexploitation of natural resources, and widespread alterations to the European natural landscape (Anderson and Grove 1987).

The nineteenth century was the era of the Big Hunt in Africa, in which European sportsmen attempted to kill almost all the game they laid eyes on. The biggest hunt in South African history, for example, took place on August 24, 1860, when Prince Alfred, Queen Victoria's sixteen-year-old son, together with Sir George Grey and a royal party, went hunting outside Bloemfontein in the Republic of the Orange Free State. During the course of the one-hour hunt, the party shot an estimated 1,000 head of game (Pringle 1982).

The establishment of hunting as a sport affected the economic livelihood of the black ethnic groups in Africa, mainly because it created the assumption that hunting as a sport was noble, whereas traditional subsistence hunting, especially with snares, was seen as less civilized. The overexploitation of wildlife in Africa led to a drastic decline in wildlife numbers, and blame for this situation in many cases was laid on the local black ethnic groups. The Europeans reacted by restricting the access of black people to wildlife, denying them legal access to weapons, making them ineligible for hunting licenses, and prohibiting their ownership of dogs (Carruthers 1993). The destructiveness of colonial hunting is exemplified by the extinction of the quagga, a zebralike animal adorned with stripes on its head, neck, shoulders, and part of its trunk. The last quagga, which was indigenous to South Africa, died in captivity in the Amsterdam Zoo in the Netherlands on August 12, 1883. Three years later the Cape colonial government extended legal protection to the quagga for the first time. However, there were none left in the veldt to protect under this new piece of legislation, as was soon discovered when the Amsterdam Zoo started looking for a replacement quagga (Pringle 1982).

Colonial hunting, though very destructive, did serve the important purpose of stimulating interest in African natural history and taxonomy, which in turn stimulated ideas about imposing some restrictions on hunting before valued species were wholly exterminated. This eventually led to the emergence of institutional and legal attempts by colonial authorities to conserve the wildlife resources on the continent. In 1900 European colonial authorities met in London in a first endeavor to discuss wildlife conservation in Africa. Before adjourning they signed a Convention for the Preservation of Animals, Birds and Fish in Africa. This was followed in 1933 by another intergovernmental conference on African wildlife, during which the colonial authorities agreed to address the threat to African wildlife through the creation of national parks and game reserves. By that time several game reserves had already been established in the colonies, such as the Umfolozi, Hluhluwe, St. Lucia, and Umdhletse game reserves in the Natal colony (1897), while the world-renowned Kruger National Park obtained national park status in the Union of South Africa in 1926. Several other national parks were created after World War II, such as the Nairobi National Park in Kenya (1946); Wankie in Southern Rhodesia (now Zimbabwe), 1951; the Serengeti in Tanganyika (now Tanzania), 1951; and the Murchison Falls and Queen Elizabeth national parks in Uganda in 1952 (Adams 2001).

The establishment of game reserves and national parks added wildlife protection to the colonial environmental agenda, which prior to that time had been primarily focused on forest conservation (in order to ensure the continued and sustained exploitation of forest resources upon which many colonial governments depended for revenue); soil conservation measures (especially in southern Africa following the 1933 drought, and in North and West Africa where the possible expansion of the Sahara desert dominated discussions and conservation activities); and water conservation measures in the more arid regions of the continent. In most cases these measures rested on the assumption of European intellectual and scientific superiority, for they emphasized changing conservation and consumption patterns among African communities through compulsory conservation schemes.

The European domination of conservation measures during the colonial era impeded the development and growth of a modern African environmental movement by alienating African people from the very natural resources that they had been conserving and protecting for centuries prior to the arrival of the European colonizers. It also imprinted the continent with the belief that conservation could succeed only when local communities were totally excluded from conservation areas. This legacy was reinforced after independence by the numerous foreign wildlife and conservation experts employed by African governments to continue conservation efforts begun during the colonial era. In addition, critics claim that the launching of the Africa Special Project in 1961 by the International Union for the Conservation of Nature (IUCN) and the UN Food and Agriculture Organization (FAO) prolonged the perception that African wildlife had to be protected from Africans (ibid.).

Despite this conservation focus of the initial years of independence, an alternative African environmental movement began to emerge shortly after independence. This emerging movement was armed with an environmental agenda that focused not on conservation but on the environmental effects of underdevelopment and poverty. Indeed, the movement "grew out of the efforts of poor, uneducated, and mostly rural people who sought to limit the environmental damage caused by inappropriate development policies that only served to worsen African living standards" (Daniels 1992). As a result, rural communities founded self-help organizations to take control of their lives and to improve their conditions. Notable groups of this type included the Harambee movement in Kenya, Ujamaa in Tanzania, the Naam Movement in Burkina Faso, and similar organizations in Senegal, Mauritania, Mali, Niger, and Togo. These self-help organizations, in which women played a central and dominating role, performed an important task in rural development throughout Africa, especially in terms of the provision of education, safe water, and health facilities, and in addressing environmental obstacles hampering development efforts. In Kenya, for example, these rural self-help organizations accounted for more than 70 percent of the 1,400 secondary schools in the country by 1990, as well as a substantial number of rural health clinics, cattle dips, water facilities, and production units. By 1991, Africa's Sahel countries (western African nations bordering the Sahara Desert) had more than 15,000 rural self-help organizations, while Kenya alone had more than 2,000 (ibid.).

Another important role-player in the African environmental movement was the Christian church, which was able to function relatively independently of the state. As a result, the church in Africa became an important vehicle for community opposition to harmful governmental policies and a central figure in various environmental and development issues. Church involvement in environmental issues in Zimbabwe, for example, led to the establishment of the Association of African Earthkeeping Churches (AAEC). This organization sought to harness traditionalist and Christian religious heritage in order to engage in sustained and meaningful environmental action (Daneel 1998).

During the 1980s the environmental movement in Africa was mostly confined to activism on local and rural levels. It differed greatly in this respect from environmental activism in the United States, which by the 1980s had become associated with highly politicized environmental activists and organizations. The main reason for the general lack of political focus of environmental activists in Africa was that the state in independent Africa remained in firm control of politics and tolerated little dissent from opposition groups. The African state therefore also remained in control of many environmental initiatives, with many nongovernmental organizations (NGOs), such as the Committees for the Defense of the Revolution in Burkina Faso, being either state initiated or state led. In many African countries the most influential environmental NGOs also had very good relationships with the government, such as the special relationship between the apartheid government in South Africa and the Habitat Council and the Wildlife Society of South Africa. As a result, criticism has been leveled at these NGOs for being complacent regarding governmental environmental abuses (Daniels 1992).

In contrast to Africa, the Middle East with its harsh environment did not capture the popular imagination of their European colonizers in the same way. As a result, scant attention was paid to environmental matters other than water conservation during the colonial era. Independence in the Middle East coincided with the emergence of petroleum as the dominant energy source in the world energy market, which ensured Middle Eastern petroleum-producing countries of a steady and increasingly profitable source of income, especially after the first oil boom that followed in the wake of the 1973 Oil Crisis and Arab oil embargo against the United States, Great Britain, and other "unfriendly" nations. Petroleum profits ensured enough money to sustain the populist and authoritarian governments of the Middle East, which entered into social contracts with their citizens in which the state undertook to provide them with the necessary development initiatives in return for popular support. Within this context, development strategies were mainly state-led, with local communities given minimal opportunity to participate in planning for their own future.

In the 1960s and 1970s the populist and authoritarian states of the Middle East tolerated little opposition from political parties, the independent press, professional bodies, independent trade unions, and NGOs, which made it difficult for civil society to organize opposition to governmental policies that were detrimental to the environment. As a result, civil society remained weak in the Middle East until the economic hardships of the 1980s convinced people to take greater control of their own futures (Bayat 2000).

The development of an environmental movement in the Middle East in the 1960s and 1970s was further undermined by the development of political Islam in the region. Political Islam found the Western—especially U.S.—roots of the modern environmental movement and activism to be unacceptable, and in direct reaction to Western preoccupation with environmental issues, political Islam identified environmental destruction as being a result of Western secular economic principles that had dominated the world economy for centuries. This position made it almost impossible to establish a direct link between misplaced economic and development priorities and the deterioration of the region's natural resource base (Kula 2001).

The Modern Environmental Movements in Africa and the Middle East

The emergence of the current highly active environmental movements in Africa and the Middle East in the 1980s was brought about by civil reaction to dismal economic and social conditions in the two regions. The affluence created by the oil windfalls of the 1970s was short-lived, and by the 1980s the majority of developing countries, including the oil exporters, found themselves in dire economic straits. A decrease in the oil price, along with a decrease in oil consumption stemming from oil and energy conservation measures implemented overseas in the late 1970s, hurt the economies of oil-exporting developing countries in both Africa and the Middle East during the 1980s. As a result, both oil-exporting and oil-importing developing countries struggled to meet the obligations of their external debt and were subjected to demands made by international financial institutions such as the International Monetary Fund and the World Bank for radical reforms to their national economies. The structural adjustment programs that went hand-in-hand with debt rescheduling and restructuring forced governments in Africa and the Middle East to open themselves up to trade liberalization, while drastic cuts were made in state subsidies for general commodities such as fuel, electricity, transport, and basic foodstuffs such as rice, sugar, and cooking oil (Bayat 2000).

These developments heightened the economic hardships experienced by people in Africa and the Middle East; they had to bear the brunt of drastic reductions in governmental spending along with the reduction in traditional subsidies that had enabled the poor to make ends meet. In the Middle East, governments withdrew from traditional social responsibilities that had characterized earlier development projects and that constituted the heart of the social contract between governments and civil society (ibid.). In Africa, the general population had profited little from the removal of colonial rule in the 1960s. Indeed, it had grown poorer during the independence period while at the same time being subjected in most countries to political systems characterized by injustice, instability, violence, and a general lack of regard for human rights. In addition, civil unrest and war continued to play havoc with both the natural and human environments in various countries. Recurring environmental problems and challenges, such as the devastating droughts in the 1980s that placed states such as Ethiopia on the front pages of international publications, further undermined the efforts of poor people to survive (Timberlake 1988).

Within this context of economic, social, and environmental hardship that prevailed in most of Africa and the Middle East by the late 1980s, there emerged, for the first time, a strong reaction from civil society against both the state and the continued deterioration of living standards and the natural environment. The emergence of a strong social response to poor political, social, economic, and environmental conditions in the two regions coincided with important changes in international politics. Of particular importance was the fall of the Berlin Wall in 1989 and the collapse of the Soviet Union shortly thereafter; these events created a new international world order that began to focus increasingly on environmental and human rights issues in the 1990s. That in turn enabled NGOs and rural communities in Africa and the Middle East to confront their governments, big business, and other institutions engaging in environmentally destructive behavior with a vigor and energy that had been absent for most of the independence era.

UN environmental initiatives, especially the convening of the Earth Summit in 1992 and the promotion of sustainable development as *the* developmental and economic blueprint that would ensure the survival of humankind on earth, also contributed greatly to the creation of a favorable climate in which rural and urban communities in Africa and the Middle East could oppose perceived environmental threats to their well-being. Another important milestone was the establishment of a new emphasis on integrated development initiatives in which the recipient communities were expected to play a central role in the planning, initialization, and sustaining of development projects.

All the above factors combined to give civil society in Africa and the Middle East a proper voice for the first time, which in turn led to the creation of a plethora of environmental organizations in the two regions, such as Earthlife Africa in South Africa, the Society for the Renewal of Nature Conservation in Liberia, ATPNE in Tunisia, Guamina in Mali, Environmental Rights Action in Nigeria, the Action Committee for the Prevention of Air Pollution in Israel, and the Galilee Society among the Palestinian community in Israel. In both Africa and the Middle East, the NGO sector has grown tremendously in the past decade, with more than 14,500 registered NGOs in Egypt, 5,000 in Tunisia, 1,586 in Lebanon, and more than 800 in Jordan by 1999. However, the development of the environmental NGO sector in the two regions still remains very dependent on the general level of political stability. Consequently, relatively stable countries such as Botswana, Burkina Faso, Kenya, Senegal, and Togo have more active environmental NGOs than do politically unstable countries such as Somalia and Sudan (Bayat 2000; Daniels 1992).

The focus areas of environmental activism in Africa and the Middle East in the past twenty years have broadened considerably to include pollution issues, health care, capacity building within local communities, corporate accountability, community conservation, biodiversity, rural and urban development, desertification, sustainable water and fisheries development, agriculture (especially organic and sustainable farming), and the training and promotion of women. Since the 1980s, environmental NGOs in Africa and the Middle East have benefited greatly from their interaction and co-operation with international environmental NGOs such as Greenpeace, Friends of the Earth, the

Rainforest Action Network, the World Wide Fund for Nature, and the World Conservation Union-IUCN. These international organizations have helped NGOs and communities in the two regions by providing research, organizational, and funding assistance as well as much-needed links to international networks so as to increase the possibility of success. For example, the struggle of the Ogoni people in Nigeria against oil production by Royal Dutch/Shell and the Nigerian government in their traditional homeland in the 1990s was publicized worldwide through the efforts of both their spokesperson, Ken Saro-Wiwa, and their strategic alliances with a variety of NGOs such as Greenpeace, the World Rain Forest Action Group, Amnesty International, Earthlife Africa, and the British retailer the Body Shop (Haynes 1999).

National and local NGOs in the two regions have also benefited from international networking and interaction by becoming national chapters/branches of well-known diversified international environmental NGOs. Both the World Wide Fund for Nature (WWF) and the IUCN have a long history of activities in Africa and the Middle East, but it is largely the expansion of the Friends of the Earth network into the two regions that has given local communities and NGOs the motivation to start addressing environmental issues from a stronger activist position. Currently, Friends of the Earth are active in nine African countries (Sierra Leone, Ghana, Togo, Benin, Nigeria, Cameroon, Tunisia, Mali, and South Africa), and Friends of the Earth Middle East works not only toward improving the Middle Eastern environment but also toward improving political relationships in this volatile region of the world. By contrast, Greenpeace is currently not represented in sub-Saharan Africa, while North Africa and the Middle East are part of the Greenpeace Mediterranean group.

Major Issues in Environmental Activism

Over the past two decades African and Middle Eastern environmental activists have been engaged in a wide range of struggles against perceived or real environmental threats. What follows is a brief discussion of some of these struggles in terms of urban development, activism against industries, activism against petroleum developments and production, community involvement in nature conservation, greening and community development projects, and traditional wildlife conservation.

Urban Development

In both Africa and the Middle East, rapid urbanization in the past few decades has brought about many environmental challenges that city dwellers and the government have to face on a daily basis. This phenomenon, fed by heavy rates of migration from rural areas to urban centers, has exceeded all expectations of city planners in the two regions. In addition, financial and institutional resources diverted to address this state of affairs have often been grossly inadequate. As a result, urban communities more often than not experience a lack of much-needed services, while acute housing shortages and impoverishment have resulted in extensive squatter settlements on the borders of the big towns and cities, accompanied by their associated social and environmental challenges.

In most African and Middle Eastern cities, basic services to poorer communities are either nonexistent or badly maintained; this has necessitated community mobilization and organization to achieve particular environmental and social objectives. In Cairo, Egypt, for example, numerous examples exist of local cooperation to improve street cleanliness, install street lighting, collect garbage, and organize sewage removal on a local basis. The community of Dar al-Salam, for example, organized a communal workforce to level streets and plant trees after the installation of a new sewerage system to ensure that flooding by sewage no longer threatened the community. Similarly, the community of Sayyida Zeinab created a minipark by simply planting trees, thereby improving the aesthetic value of their urban community as well as the quality of their local air (Hopkins and Mehanna 1996).

The Galilee Society in Israel is a joint venture between Jews and Palestinians that works toward the protection of the environmental rights of the Arab minority in the country. The society not only organizes much-needed sewage collection and treatment systems in Palestinian communities but also has an impressive list of achievements against local industrial polluters. In June 2000, for example, the society reached an agreement with the owners of the Sasa Chemia Factory that called for the closure of the factory so that clean-up of waste produced by the factory could take place. It further succeeded in the same year in closing down the Kinneret Stone Quarry, which was a major source of air pollution in the Nazareth District (Galilee Society 2000).

During the apartheid era in South Africa (1948–1994), little town planning went into black urban communities that bordered the white towns and cities. Consequently, these communities often lacked adequate housing and infrastructure and services such as waste removal, clean water, and sewage collection. As a result, many local black communities organized themselves into community-based NGOs such as the National Environmental Awareness Campaign (NEAC), founded in Soweto in 1978 by Japhta Lekgheto, to improve their urban environment. From the start NEAC set out to promote environmental awareness, launched a campaign to combat the lack of waste removal services in Soweto (Operation Clean Up), and set up a recreational center in Dobsonville Park to help provide for the social needs of area youth. NEAC was also one of the first NGOs in South Africa to link the apartheid system with the dismal conditions in black townships. "Blacks have always had to live in an environment that was neither beautiful nor clean," said Lekgheto. "We have not had proper housing, roads or services because the authorities would not accept that we were a permanent part of the city scene" (Cock 1990).

Activism against High-Polluting Industries

Industrial development in Africa and the Middle East in the second half of the twentieth century went hand in hand with widespread environmental problems such as air, water, and soil pollution, as well as the overexploitation of natural resources. This state of affairs was caused mainly by a lack of governmental attention to environmental issues prior to the 1990s, the lack of enforceable environmental standards, and the unwillingness of government to enforce existing environmental legislation for fear that it would curb industrial development seen in both regions as vital to efforts to diversify their economies (Marzouk and Kaboudan 1989).

When governments have proven themselves unwilling to act to curb industrial pollution, the local communities close to these industries have had to bear the brunt of bad environmental practices. This situation has led to numerous struggles against industrial developments in Africa and the Middle East. In South Africa, for example, the residents of Steel Valley in Vanderbijl Park sued the steel giant Iscor in 2001 for polluting both the ground and water resources for three decades. During this time many residents became ill because of the industrial pollution; they lost their only sources of water because of widespread groundwater pollution and had their soil polluted to such an extent that very few of the small farms can still be utilized (Magardie 2001).

In Cairo, residents of Kafr al-Elow started to organize in the late 1980s against a local cement plant that refused to install filters to reduce its dust pollution of the community. Community activists mobilized support from the news media and tried in vain to lobby government officials to support their cause, while a local lawyer has been trying since 1988 to use court cases to force the cement plant to install filters. Unfortunately for the residents of Kafr al-Elow, their struggle still continues (Hopkins and Mehanna 1996). The residents of Romi River Valley in northern Nigeria, on the other hand, were much more successful in combating large-scale pollution of the Romi River and Valley by effluents from the Kaduna oil refinery. In 1989 they succeeded in convincing the Nigerian federal and Kaduna state governments to clean up the pollution (Areola 1998).

In Jordan a unique network called the Jordanian Network for Environmentally Friendly Industries (JNEFI), developed in response to widespread environmental pollution, came into existence in the 1990s. This organization

Nigeria's Ogoniland has been a hotbed of environmental activism. GREENPEACE/CORBIS SYGMA

addresses environmental problems in industries by identifying specific problems and concerns inside industrial facilities and finding ways to treat and minimize these environmental problems (Jordanian Network for Environmentally Friendly Industries 2003).

Activism against Petroleum Development and Production

Since the early 1990s major petroleum companies have been challenged by local communities adjacent to oil production sites in many parts of the developing world, including a number of African nations. Tired of decades of environmental and economic marginalization, these communities publicized their struggles to stop environmental abuses associated with energy exploration in efforts to enlist international support for their causes.

The activities of Ken Saro-Wiwa and the Ogoni people against Shell Nigeria and the Nigerian federal government in the first half of the 1990s serve as a good example of environmental activism against the petroleum industry. The Ogoni people live in the Niger Delta in eastern Nigeria where they subsist on farming and fishing. Oil was first discovered in Ogoniland in 1958 by Shell-BP, and production started in the same year and continued until 1993. During this period petroleum became a dominating aspect of life in the Ogoni territory. But the Ogoni never reaped the economic benefits of petroleum production in their territory and had to bear the environmental consequences of bad environmental practices in the form of continuous noise pollution, air pollution as a result of gas flaring, and widespread water and soil pollution by crude oil—all of which harmed their local economy, health, and standard of living. After decades of environmental and economic marginalization, the Ogoni organized themselves in 1990 into the Movement for the Survival of the Ogoni People (MOSOP), of which the Ogoni playwright Ken Saro-Wiwa became the most vocal spokesperson. After waging a three-year struggle against Shell Nigeria, the Ogoni succeeded in terminating petroleum production in their traditional territory when Shell Nigeria withdrew indefinitely from Ogoniland in January 1993. The struggle between the Ogoni and the federal government, however, continued and reached a peak in November 1995 with the execution of Saro-Wiwa and eight other Ogoni for their alleged involvement in the murder of four Ogoni elders during an uprising in Goikoo in 1994 (Haynes 1999).

The Ogoni struggle has motivated numerous communities with similar grievances to take up action against the petroleum industry. It has also provided them with a "blueprint" on how to articulate their grievances to both the federal government and the oil companies that operate in their territory. A variety of different ethnic minority groups, such as the Ijaw, through the Movement for the Survival of the Izon (Ijaw) Ethnic Nationality, have subsequently embarked on campaigns based upon that of MOSOP.

In neighboring Cameroon, meanwhile, fierce environmental opposition emerged in the 1990s to plans to build an oil pipeline between the Doba oil fields in Chad and coastal Cameroon. This opposition extended to the World Bank, which faced strong international opposition for providing the financing for the project. According to opponents, the Chad-Cameroon Oil and Pipeline Project presents numerous environmental problems, not least of which is that the 960-kilometer oil pipeline will be raised above ground instead of installed beneath the surface and will traverse ecologically sensitive areas. In addition, the indigenous Bakka and Bakola "pygmy" groups will be uprooted from their traditional territory in the rain forest region of southern Cameroon. As coverage of the controversial project intensified, the Bakka and Bakola received widespread support from environmental and human rights groups, both in Cameroon and outside the country. The World Bank subsequently commissioned a new environmental impact assessment for the whole project. Although this campaign has succeeded in convincing two of the original oil company partners, namely Shell and ELF (France), to withdraw from the project, their places were quickly taken by Petronas (Malaysia) and Chevron, which, together with the main partner in the project, ExxonMobil, started with the construction of the pipeline in 2000 (Uriz 2001).

Although the environmental activism of the Ogoni and that surrounding the Chad-Cameroon Oil and Pipeline Project are well known, the struggle of

Ken Saro-Wiwa, environmental and human rights activist. GREENPEACE/CORBIS SYGMA

Nigeria, Oil, and the Case of Ken Saro-Wiwa

Kenule Beeson Saro-Wiwa was born in 1941 in the Nigerian settlement of Bori. The son of a tribal chief of the Ogoni, one of the nation's indigenous peoples, Saro-Wiwa emerged as a spokesman for disenfranchised Ogoni at a young age. During the 1970s and 1980s, he and other Nigerians became convinced that the military dictators running the country were exporting the nation's oil reserves exclusively for their own gain, instead of reinvesting the earnings in projects to provide electricity, running water, and medical care for ordinary citizens. Saro-Wiwa and his allies asserted that the Ogoni people were suffering particularly inequitable treatment. They noted that much of the oil and natural gas that was being extracted from Nigerian soil was located in "Ogoniland," an area of the Niger Delta in southeastern Nigeria where generations of Ogoni people had maintained subsistence lifestyles through fishing and farming. But the profits generated by these operations were divided between the Nigerian government and Shell and other foreign oil companies, with almost no revenue trickling down to Ogoni communities. This situation was further exacerbated by the fact that oil operations, unencumbered by environmental regulations, were degrading forests, fields, and rivers throughout Ogoniland. The loss of previously productive streams and agricultural lands took a heavy toll on many Ogoni households and kindled growing resentment and anger across the delta.

By 1990, Saro-Wiwa's career path had taken him from teaching to writing for radio, television, and theatre productions. But in October of that year, Nigerian police crushed a peaceful protest staged by one Ogoni community against the ecological toll of oil operations in the area. In the wake of the savage assault, which resulted in the deaths of eighty villagers and the destruction of nearly 500 homes, Saro-Wiwa and several other Ogoni leaders formed the Movement for the Survival of Ogoni People (MOSOP). The group immediately called for fundamental changes in operations by oil companies operating in the Niger Delta, as well as billions of dollars in financial reparations.

By 1993 an estimated 300,000 people had pledged their support to MOSOP and Saro-Wiwa, the organization's leading spokesman. But while MOSOP emphasized nonviolent means of protest—and carried out many peaceful rallies throughout 1993 and 1994—some disaffected Ogoni began to use violence and sabotage against Shell's delta operations. Alarmed by the turn of events, Shell announced a temporary closure of its Ogoniland facilities in late 1993. This in turn prompted Nigeria's military junta to strip the country's citizens of basic liberties, dramatically curtail press

(continues)

freedoms, and strike back against Ogoni protestors with armed troops pulled from rival ethnic clans. Over the next several months, international human rights organizations reported numerous incidents of murder, rape, torture, and other human rights violations in Ogoniland.

During this period, Saro-Wiwa struggled with only limited success to keep MOSOP's various elements, which ranged from angry youth to traditional chieftains, unified. In the meantime, an atmosphere of menace descended over much of the delta, fueled by proliferating rumors and reports that the government intended to assassinate Saro-Wiwa and other activists. In May 1994, the growing tensions between MOSOP's young radicals and the tribal chieftains exploded into violence, as several chiefs that had decided to sever their ties with the organization were murdered in the village of Giokoo. The following day, government troops descended en masse on Ogoniland, terrorizing villages and arresting Saro-Wiwa and more than a dozen other MOSOP leaders.

In 1995, Nigerian authorities formally charged Saro-Wiwa with inciting the killings at Giokoo. During the trial, evidence that the verdict against Saro-Wiwa was preordained became so overwhelming that several of the activists' attorneys resigned in protest. At the trial's conclusion, Saro-Wiwa and eight other MOSOP activists were found guilty and sentenced to be executed. Saro-Wiwa quickly issued a statement in which he declared that he was "a man of peace" who had been driven to act because he had become "appalled by the denigrating poverty of my people who live on a richly endowed land, distressed by their political marginalization and economic strangulation, and angered by the devastation of their land." He added that he remained proud of his efforts to "preserve [the Ogonis'] right to life and to a decent living" and his push to "usher to this country as a whole a fair and just democratic system which protects everyone and every ethnic group."

Saro-Wiwa's death sentence drew worldwide condemnation from environmental and human rights groups, and numerous international governments urged Nigeria to commute or adjust the sentences. But on November 10, 1995, Saro-Wiwa and the other convicted MOSOP leaders were executed by hanging. Since that time, Nigerian authorities have maintained a heavy security presence in Ogoniland. The government and Shell Oil have also increased funding for schools and hospitals in the region, and new environmental protection measures have been implemented for some drilling and production operations. Nonetheless, environmental organizations contend that ecological conditions in extensive parts of Ogoniland are grim. In the United States, meanwhile, a federal judge ruled in late 2002 that Shell must face charges brought by surviving relatives of Saro-Wiwa that the company was complicit in his detention, trial, and hanging.

(continues)

Sources:

Daniels, Anthony. 2000. "The Perils of Activism: Ken Saro-Wiwa." *New Criterion* 18 (January).

Hammer, Joshua. 1996. "Nigeria Crude: A Hanged Man and an Oil-Fouled Landscape." *Harper's Magazine* 292 (June).

Sachs, Aaron. 1995. *Eco-Justice: Linking*

Human Rights and the Environment. Worldwatch Paper 127. Washington, DC: Worldwatch.

Saro-Wiwa, Ken. 1992. *Genocide in Nigeria: The Ogoni Tragedy.* Port Harcourt, Nigeria: Saros International.

————. 1996. "Final Statement to the Tribunal." *Social Justice* 23 (winter).

environmental activists in Zimbabwe against oil exploration remains largely unrecognized. In 1990 environmentalists in Zimbabwe successfully blocked Mobil (now ExxonMobil) from conducting oil exploration in a national park in the Zambezi Valley without an environmental impact assessment. This action only temporarily halted oil exploration, but it did ensure that oil exploration in the Zambezi Valley would be undertaken in a more environmentally sensitive manner (Daniels 1992).

Community Involvement in Nature Conservation

As mentioned earlier, the conservation of Africa's natural environment started in the colonial era when colonial governments created game reserves and national parks to protect wildlife on the continent. Colonial conservation initiatives rested on the assumption that people (especially black people) were the enemies of conservation and that they should be kept out of protected areas. It therefore became accepted practice that people should "make room" for protected areas, which normally meant the forced removal of communities living within the proposed borders. Further, local communities were not allowed to utilize the resources within the protected areas, which impacted severely on their traditional economic modes of production. This perspective of nature conservation continued well into the independent era, with local communities rarely benefiting from the protected areas, which existed mainly for the benefit of foreign and local tourists.

The traditional conservation practices that excluded communities from participation in nature conservation were challenged in the 1980s by a number of projects that helped to shift the conservation ethos away from alienating communities from their traditional land and resources toward a conservation ethos that allowed for the direct participation of neighboring communities in protected areas. The first such initiatives started in Purros in Namibia in the mid-1980s when Garth Owen-Smith and Margaret Jacobsohn set up a project on behalf of a South African NGO, the Endangered Wildlife Trust (EWT). In Purros, a village in Namibia, Owen-Smith and Jacobsohn succeeded in involving the whole community in wildlife conservation and set up mechanisms to ensure that the community benefited from the tourists that visit the area. In light of the successes of the Purros project, the EWT hosted an international symposium on national parks, nature reserves, and neighboring communities in 1988. This symposium provided the first forum ever at which communities near and within protected areas could voice their opinions regarding the use of these areas (Koch, et al. 1990).

A similar project was set up in Zimbabwe with the creation of the Zimbabwe Communal Areas Management Program for Indigenous Resources (CAMP-FIRE) in 1989. CAMPFIRE, which is a collaborative effort by the Department of National Parks and Wild Life, the Zimbabwe Trust, the University of Zimbabwe, and the World Wide Fund for Nature, works toward the development of an effective wildlife management system that balances public, private, and community interests. It encourages rural communities to determine how their land and wildlife will be used according to their own needs and priorities, while communities use the proceeds from tourism, hunting, and, at times, ivory sales to build schools, clinics, and other needed facilities (Metcalfe 1999).

Both CAMPFIRE and the Purros project have succeeded in involving local people in conservation efforts in ways that are beneficial to both the natural environment and the local communities. These efforts have been so successful that a number of African countries, in particular Zambia, Tanzania, and Mozambique, have begun to make use of this model. In South Africa, community involvement in conservation became a norm during the course of the 1990s. At that time the government officially pursued a policy of community involvement in nature conservation to the benefit of a diversity of communities, including those in Kosi Bay and those living in and on the borders of the well-known Kruger National Park. In the Richtersveld in the Northern Cape, the new conservation ethos enabled the Richtersveld Community Committee to successfully negotiate grazing rights and management participation in the proposed Richtersveld National Park with the National Parks Board (Fig 1991; Boonzaier 1991).

Greening and Community Development Projects

Throughout the independence era greening and community-based development projects proved a very popular and effective way for environmental

Wangari Maathai, Founder of the Green Belt Movement

Wangari Maathai, leader of the Green Belt Movement in Kenya and advocate of environmental and women's issues. ADRIAN ARBIB/CORBIS

Wangari Maathai is one of Africa's most famous advocates for the environment and human rights. For the better part of three decades, she has worked tirelessly to protect threatened forests in Kenya and elsewhere, and her Green Belt Movement, founded in the late 1970s, has been credited with planting tens of millions of trees in Kenya and around the world.

Born in 1940 in a farming village in south-central Kenya, Wangari Muta Maathai attended college in the United States on academic scholarships. She returned to Kenya in 1966, armed with a master's degree in anatomy and a determination to use her schooling in service to her country. In 1971 she became the first woman to earn a Ph.D. from the University of Nairobi, and over

(continues)

the course of the decade she became the university's first woman lecturer, professor, and department head.

During the mid-1970s Maathai's research convinced her that deforestation across Kenya was taking an unacceptable toll on biodiversity, depriving rivers of a vital moisture catchment resource, and reducing soil quality that was essential for agricultural productivity. In 1977, Maathai enlisted the membership of the National Council of Women of Kenya in a tree planting program. "We started with seven trees in a small park in Nairobi," she recalled. "We had no nursery, no staff, and no funds, only a conviction that there was a role for ordinary country people in efforts to solve environmental problems." This modest initiative turned out to be the earliest stirring of the Green Belt Movement.

In subsequent months, increasing numbers of volunteers—mostly women—turned out to help the program. By the early 1980s, Maathai and the rapidly expanding Green Belt Movement were garnering international attention. The organization used tree planting not only as a vehicle for environmental restoration and topsoil preservation but also as the cornerstone of education programs touting the economic, ecological, and aesthetic benefits of sustainable land management. During the late 1980s and early 1990s, Maathai's efforts in behalf of the environment and the subsistence farmers of Kenya were officially recognized by organizations ranging from the Better World Society, which gave her the 1986 Award for the Protection of the Global Environment, to the UN's Africa Prize for Leadership in 1991. Other honors received by Maathai included the Windstar Award for the Environment (1988) and the Goldman Environmental Prize (1991). She also helped found the Women's Environment and Development Organization, and was a featured speaker at the World Conference on Women in 1995.

But Maathai's outspoken defense of Kenya's natural forests and her habitat restoration schemes have not been universally popular within her own country. In 1989, for example, she ran afoul of Kenyan president Daniel arap Moi when she organized an effective public campaign against the planned building of a skyscraper in the middle of Nairobi's downtown Uhuru Park. Maathai and other detractors condemned the proposed building as an expensive boondoggle that would eliminate one of the few green spaces left in the city. She was subsequently vilified by politicians in Moi's ruling party—and by Moi himself, who charged that it was "un-African and unimaginable for a woman to challenge or oppose men." But the protests were so effective that foreign investors withdrew from the project and the trees of Uhuru Park were left unscathed.

In 1991, Maathai helped start an opposition group called the Forum for the Restoration of Democracy that was opposed to the one-party dictatorship

(continues)

that had ruled Kenya since 1982. Over the next few years, she endured tear gas, physical attacks from police that rendered her unconscious, threats on her life, and repeated arrests. But she remained staunchly devoted to the causes of civil rights, women's rights, and forest conservation throughout the 1990s.

At the close of the twentieth century, meanwhile, the Green Belt Movement that Maathai helped launch was still thriving. Assisted by an estimated 6,000 women's groups worldwide, the movement has reforested large tracts of cleared land in Kenya and elsewhere, and it has repeatedly gone to court to block government-sanctioned timber concessions in species-rich forests.

In December 2002, Maathai won a seat in Kenya's parliament as a member of the country's National Rainbow Coalition (NARC), a new ruling party that has pledged to enact policies that will safeguard forests. After learning of her victory, she quickly vowed to focus on preserving Kenya's remaining natural forests from logging. "I have been trying to change policy from outside, and now I will have the opportunity to change policy from inside," she stated.

Sources:

Brill, Alida, ed. 1995. *A Rising Public Voice: Women in Politics Worldwide.* New York: Feminist Press at CUNY.
Maathai, Wangari. 2003. *The Green Belt Movement: Sharing the Approach and the Experience.* Rev. ed. New York: Lantern.

activists to address environmental and development issues to the advantage of local communities. The best known of these greening projects is Kenya's Green Belt Movement. This antideforestation project was founded in 1977 by Nairobi University professor Wangari Maathai. It was based on the principle of empowerment, for it enlisted the support of rural women who had to travel greater distances to collect firewood as forests retreat because of overexploitation. The Green Belt Movement distributes tree saplings among rural women and then pays them for each tree that survives. In this way, women not only earn an independent income but also plant enough firewood to ensure the survival of more valuable tree species in the forested areas and directly counter deforestation and its associated environmental problems. Green Belt has been an enormous success in Kenya, with more than 20 million trees planted to date (Haynes 1999).

The greening of some black and colored communities in South Africa also began in earnest in the 1980s with the founding of Abalimi Bezekhaya ("Planters of the Home") by a Catholic welfare and developmental organization. Initially, Abalimi Bezekhaya focused on attempts to stimulate and promote an organic food garden culture among black communities in the greater Cape Town area in order to help people produce their own food. During the course of the 1980s it broadened its agenda to include tree planting, general greening of townships, and environmental education. Two garden centers were set up in Nyanga (1985) and Khayelitsha (1989) to provide a low-cost service to township residents ("Abalimi Bezekhaya" 1992). Similar projects were launched by the Africa Tree Centre (1984, in Edendale), Natsoc (1984, in the Cape Flats), Ecolink (1985, in Gazankulu, KaNgwane, and Lebowa), and Khanyisa (1988, in Langa, Guguletu, and Khayelitsha).

Community development projects aimed especially at improving the lives of rural women are also a common focus of the environmental movement in Africa and the Middle East. In South Africa, for example, the Mboza Village Project in northern KwaZulu was established in the early 1980s to help develop the community and to create job opportunities. It began as a sewing project for women in the area and as a literacy center but expanded to incorporate issues such as the provision of safe water and primary health care (Cooper 1994). Similar projects exist in most countries in the two regions, such as in Nigeria (Environmental Rights Action), Togo (Friends of the Earth Togo), Sierra Leone (Friends of the Earth Sierra Leone), and in Mali (Guamina).

Traditional Wildlife Conservation

For many of the highly politicized environmental activists, traditional wildlife conservation does not merit much attention, since it fails to address the "real" environmental issues such as pollution and environmental abuse. But, in terms of the wildlife heritage of both Africa and the Middle East, traditional conservation NGOs do play an important role in the conservation of an integral part of the African and Middle Eastern natural environment. Wildlife conservation societies, such as the Royal Society for the Conservation of Nature in Jordan, the Society for the Protection of Nature in Israel, the Nigerian Nature Conservation Society, and the Wildlife and Environment Society of South Africa, exist in most countries, and some of these organizations have done remarkable work to protect wildlife and natural resources. In most cases these organizations work in close collaboration with international wildlife and nature conservation NGOs such as the World Conservation Union-IUCN and the WWF to counter habitat destruction and species extinction. The efforts of these organizations have helped bring about conservation success stories such as the successful reintroduction of the Arabian oryx in Jordan in the late 1970s, the international protection of elephants and rhinoceros, and the very successful breeding program in South

Africa to ensure the survival of the black rhinoceros, which was taken from the brink of extinction to its current thriving population in southern Africa.

In South Africa, the 1990s also saw the emergence of radical proanimal environmental activists whose activities added a new dimension to wildlife conservation on the African continent. The oldest example of the sustained utilization of wildlife resources in the country was the annual culling of the Cape fur seal population to control their numbers as well as to provide an income for coastal communities. This annual seal culling had always been a highly controversial and emotive issue, but it continued for more than 300 years until 1990, when the Minister of Environmental Affairs announced the indefinite postponement of the culling of 30,000 seals at Kleinsee. Even though anti–animal cruelty NGOs had campaigned for years against seal harvesting, credit for the postponement belonged to Earthlife Africa, the Seal Action Group, the World Society for the Protection of Animals, Save our Seals, and the Front for Animal Liberation and Conservation of Nature. Their highly emotional (and at times violent) campaign led to the appointment of a committee to investigate the scientific aspects of sealing in August 1990. Although the committee found no scientific reasons for halting the proposed harvesting, and further recommended the controlling of the Cape fur seal population at Kleinsee, the South African government decided to temporarily suspend all commercial seal harvesting in South African waters in February 1991 (Republic of South Africa 1990). This suspension was still in place in 2002.

Environmental Activism as Part of the Middle Eastern Peace Initiative

For most environmentalists, environmental activism serves the purpose of addressing environmental ills in society. In the volatile Middle East, however, it also performs an important function in bridging sharp differences between the various countries and the Jewish and Arabic communities. It thus serves to promote peace initiatives in the region by fostering cooperation between parties that are in serious conflict about other issues. These initiatives promote integrated regional approaches to environmental issues because the people and countries of the region are dependent on many of the same natural resources for their survival. In recognition of this reality, the international Middle East Peace Process includes two working groups concerned with environment-related issues. One group addresses environmental management, maritime pollution, desertification, water quality, sewage and waste management, and hazardous waste, and another working group focuses on water management practices, quality, and conservation, as well as regional water management and cooperation.

Friends of the Earth International also contributed to environmental cooperation in the Middle East with the founding of EcoPeace in 1994 (it became Friends of the Earth Middle East in 1998), which brought environmentalists from Egypt, Israel, Jordan, and Palestine together to work toward improving the shared resources in the region (Friends of the Earth Middle East 2003). Middle Eastern and North African countries further work together on a number of international initiatives to protect the Mediterranean, such as MED Forum (Forum of Mediterranean NGOs for Ecology and Sustainable Development) and Greenpeace Mediterranean, which fosters cooperation by emphasizing the importance of bridging political differences in order to improve and protect common natural resources (MED Forum; Greenpeace Mediterranean).

Sources:

1992. "Abalimi Bezekhaya: The People's Garden Center." *South African Outlook* (June).

Adams, W. M. 2001. *Green Development: Environment and Sustainability in the Third World.* London: Routledge.

Anderson, David, and Richard Grove. 1987. *Conservation in Africa: People, Policies and Practice.* Cambridge: University of Cambridge Press.

Areola, O. 1998. "Comparative Environmental Issues and Policies in Nigeria." In U. Desai, ed., *Ecological Policy and Politics in Developing Countries: Economic Growth, Democracy, and Environment.* Albany: State University of New York Press.

Bayat, Asef. 2000. "Social Movements, Activism and Social Development in the Middle East." *Civil Society and Social Movements Programme Paper no. 3, November 2000.* Geneva: UN Research Institute for Social Development.

Boonzaier, E. 1991. "People, Parks and Politics." In M. Ramphele and C. McDowell, eds., *Restoring the Land: Environment and Change in Post-Apartheid South Africa.* London: Panos.

Carruthers, E. J. 1993. "'Police Boys' and Poachers: Africans, Wildlife Protection and National Parks, the Transvaal 1902 to 1950." *Koedoe* 36.

Cock, J. 1990. "Ozone-Friendly Politics." *Work in Progress* (May).

Cooper, Carole. 1994. "People, the Environment and Change." *South African Institute of Race Relations Spotlight* (October).

Daneel, M. L. 1998. *African Earthkeepers 1: Interfaith Mission in Earth-Care.* Pretoria: Unisa.

Daniels, Nomsa. 1992. *Protecting the African Environment: Reconciling North-South Perspectives.* Washington, DC: Sidney Kramer.

Fig, D. 1991. "Flowers in the Desert: Community Struggles in Namaqualand." In J. Cock and E. Koch, eds., *Going Green: People, Politics and the Environment in South Africa.* Cape Town: Oxford University Press.

Friends of the Earth Middle East. *History of Friends of the Earth Middle East (FoEME).* Available at www.foeme.org/main/about.htm (accessed January 2003).

The Galilee Society. 2000. *Annual Report 2000*. Available at www.gal-soc.org/2000_7.html (accessed January 2003).

Greenpeace Mediterranean. *About Us*. Available at www.greenpeacemed.org.mt/about.html (accessed January 2003).

Haynes, Jeff. 1999. "Power, Politics and Environmental Movements in the Third World." *Environmental Politics* 8 (spring).

Hopkins, Nicholas, and Sohair Mehanna. 1996. "Pollution, Popular Perceptions and Grassroots Environmental Activism." *Middle East Report*, no. 202 (winter).

Jordanian Network for Environmentally Friendly Industries. *About JNEFI*. Available at www.jnefi.foe.org.jo/About_Jnfe/about_jnfe.html (accessed January 2003).

Koch, E., D. Cooper, and H. Coetzee. 1990. *Water, Waste and Wildlife: The Politics of Ecology in South Africa*. Johannesburg: Penguin.

Kula, E. 2001. "Islam and Environmental Conservation." *Environmental Conservation* 28.

Magardie, Khadija. 2001. "Vaal Community Being Poisoned to Death." *Mail and Guardian* (February 9).

Marzouk, M. S., and Mahmoud A. Kaboudan. 1989. "A Retrospective Evaluation of Environmental Protection Projects in Kuwait." *Journal of Developing Areas* 23 (July).

MED Forum. *What Is MED Forum?* Available at www.pangea.org/medforum/quees/quees_en.htm (accessed January 2003).

Metcalfe, Simon. 1999. "The Zimbabwe Communal Areas Management Program for Indigenous Resources (CAMPFIRE)." *Exploring Our Future* (winter).

Pringle, John A. 1982. *The Conservationists and the Killers*. Cape Town: T. V. Bulpin and Books of Africa.

Republic of South Africa. 1990. *Report of the Subcommittee of the Sea Fisheries Advisory Committee Appointed at the Request of the Minister of Environment Affairs and of Water Affairs, to Advise the Minister on Scientific Aspects of Sealing*. Cape Town: Government Printer.

Timberlake, Lloyd. 1988. *Africa in Crisis: The Causes, the Cures of Environmental Bankruptcy*. London: Earthscan.

United States Embassy, Jordan. *MEPP: A User's Guide to the Multilateral Bodies of the Middle East Peace Process*. Available at www.usembassy-amman.org.jo/Enviro/MEPP2.html (accessed January 2003).

Uriz, Genoveva H. 2001. "To Lend or Not to Lend?: Oil, Human Rights, and the World Bank's Internal Contradictions." *Harvard Human Rights Journal* 14 (spring).

Appendix

INTERNATIONAL ENVIRONMENTAL AND
DEVELOPMENTAL AGENCIES, ORGANIZATIONS, AND PROGRAMS
ON THE WORLD WIDE WEB

African-Eurasian Migratory Waterbird
Agreement (AEWA)
http://www.unep-wcmc.org/
AEWA/index2.html

Albertine Rift Conservation
Society (ARCOS)
http://www.unep-wcmc.org/arcos/

Association of Southeast
Asian Nations (ASEAN)
http://www.asean.or.id/

Biodiversity Planning Support
Programme (BPSP)
http://www.undp.org/bpsp/

BirdLife International (BI)
http://www.birdlife.net

Botanic Gardens Conservation
International (BGCI)
http://www.bgci.org.uk/

CAB International (CABI)
http://www.cabi.org/

Centre for International
Forestry Research (CIFOR)
http://www.cifor.org/

Circumpolar Protected Areas
Network (CPAN)
http://www.grida.no/caff/
cpanstratplan.htm

Commission for Environment
Cooperation (CEC) (North
American Agreement on
Environmental Cooperation)
http://www.cec.org/

Commission on Genetic Resources
for Food and Agriculture (CGRFA)
http://www.fao.org/ag/cgrfa/
default.htm

Commission for Sustainable
Development (CSD)
http://www.un.org/esa/sustdev/csd.htm

Committee on Trade and Environment
(CTE), World Trade Organization
http://www.wto.org/english/
tratop_e/envir_e/envir_e.htm

Conservation International (CI)
http://www.conservation.org/

Consultative Group on International
Agricultural Research (CGIAR)
http://www.cgiar.org/

Convention on Biological
Diversity (CBD)
http://www.biodiv.org/

Convention on International Trade in
Endangered Species of Wild Fauna
and Flora (CITES)
http://www.cites.org/

Convention on Migratory
Species of Wild Animals (CMS)
http://www.unep-wcmc.org/cms

European Centre for Nature
Conservation (ECNC)
http://www.ecnc.nl/

European Community (EC)
http://europa.eu.int/

European Environment
Agency (EEA)
http://www.eea.eu.int/

Forest Stewardship Council (FSC)
http://www.fscoax.org/index.html

Foundation for International
Environmental Law and
Development (FIELD)
http://www.field.org.uk/

Global Assessment of Soil
Degradation (GLASOD)
http://www.gsf.de/UNEP/glasod.html

Global Biodiversity
Information Facility (GBIF)
http://www.gbif.org

Global Coral Reef
Monitoring Network (GCRMN)
http://coral.aoml.noaa.gov/gcrmn/

Global Forest Resources Assessment
2000 (FRA 2000), UN Food and
Agriculture Organization
http://www.fao.org/forestry/fo/fra/
index.jsp

Global International Waters Assessment
(GIWA), UN Environment Programme
http://www.giwa.net/

Global Invasive Species
Programme (GISP)
http://globalecology.stanford.edu/DGE/
Gisp/index.html

Global Resource Information Database
(GRID), UN Environment Programme
http://www.grid.no

Inter-American Biodiversity
Information Network (IABIN)
http://www.iabin.org/

Intergovernmental Oceanographic
Commission (IOC), UN Educational,
Scientific, and Cultural Organization
http://ioc.unesco.org/iocweb/

Intergovernmental Panel on
Climate Change (IPCC)
http://www.ipcc.ch/index.html

International Center for Agricultural
Research in the Dry Areas (ICARDA)
http://www.icarda.cgiar.org/

International Centre for Living Aquatic
Resources Management (ICLARM)
http://www.cgiar.org/iclarm/

International Centre for Research in
Agroforestry (ICRAF)
http://www.icraf.cgiar.org/

International Cooperative
Biodiversity Groups (ICBG)
http://www.nih.gov/fic/programs/icbg.
html

International Coral Reef
Action Network (ICRAN)
http://www.icran.org

International Coral Reef
Information Network (ICRIN)
http://www.environnement.gouv.fr/
icri/index.html

International Council for the
Exploration of the Sea (ICES)
http://www.ices.dk/

International Council for Science (ICSU)
http://www.icsu.org/

International Food Policy Research
  Institute (IFPRI)
http://www.ifpri.org/

International Forum on Forests (IFF),
  Commission on Sustainable
  Development
http://www.un.org.esa/sustdev/
  forests.htm

International Fund for
  Agricultural Development (IFAD)
http://www.ifad.org/

International Geosphere-
  Biosphere Programme (IGBP)
http://www.igbp.kva.se/

International Institute of
  Tropical Agriculture (IITA)
http://www.iita.org

International Maritime
  Organization (IMO)
http://www.imo.org/

International Rivers Network (IRN)
http://www.irn.org/

International Union of
  Biological Sciences (IUBS)
http://www.iubs.org/

Man and the Biosphere Program (MAB),
  UN Educational, Scientific, and
  Cultural Organization
http://www.unesco.org/mab/index.htm

Marine Stewardship Council (MSC)
http://www.msc.org/

Organization of African
  Unity (OAU)
http://www.oau-oau.org/

Organization for
  Economic Cooperation
  and Development (OECD)
http://www.oecd.org/

Ozone Secretariat Homepage
http://www.unep.ch/ozone/

Pan-European Biological and Landscape
  Diversity Strategy (PEBLDS)
http://www.strategyguide.org/

Program for the Conservation of
  Arctic Flora and Fauna (CAFF),
  Arctic Council
http://www.grida.no/caff/

Protocol Concerning Specially
  Protected Areas and Wildlife (SPAW)
http://www.cep.unep.org/law/
  cartnut.html

Ramsar Convention on Wetlands of
  International Importance (RAMSAR)
http://www.ramsar.org/

South African Development
  Community (SADC)
http://www.sadc.int/

South Pacific Regional
  Environmental Programme (SPREP)
http://www.sprep.org.ws/

Species Survival Commission (SSC),
  World Conservation Union
http://iucn.org/themes/ssc/index.htm

TRAFFIC (the joint wildlife trade
  monitoring programme of World
  Wide Fund for Nature and World
  Conservation Union)
http://www.traffic.org

United Nations Centre for
  Human Settlements (UNCHS)
http://www.unchs.org

United Nations Children's
  Fund (UNICEF)
http://www.unicef.org

United Nations Conference on
  Environment and Development

(UNCED),
Rio de Janeiro, June 1992
http://www.un.org/esa/sustdev/
agenda21.htm

United Nations Conference on Trade
and Development (UNCTAD)
http://www.unctad.org/

United Nations Convention to Combat
Desertification (UNCCD)
http://www.unccd.int/main.php

United Nations Convention
on the Law of the Sea (UNCLOS)
http://www.un.org/Depts/los/
index.htm

United Nations Development
Programme (UNDP)
http://www.undp.org/

United Nations Educational, Scientific,
and Cultural Organization (UNESCO)
http://www.unesco.org/

United Nations Environment
Programme (UNEP)
http://www.unep.org/

United Nations Food and
Agriculture Organization (FAO)
http://www.fao.org/

United Nations Forum
on Forests (UNFF)
http://www.un.org/esa/sustdev/
forests.htm

United Nations Framework Convention
on Climate Change (UNFCCC)
http://www.unfccc.de/index.html

United Nations Industrial
Development Organization (UNIDO)
http://www.unido.org/

World Agricultural Information Centre
(WAIC), UN Food and Agriculture

Organization
http://www.fao.org/waicent/search/
default.htm

World Bank (WB)
http://www.worldbank.org

World Commission
on Dams (WCD)
http://www.dams.org/

World Commission on Protected Areas
(WCPA), World Conservation Union
http://www.wcpa.iucn.org/

World Conservation
Monitoring Centre (WCMC)
http://www.unep-wcmc.org

World Conservation Union (IUCN)
http://www.iucn.org/

World Health Organization (WHO)
http://www.who.int

World Heritage Convention (WHC)
http://www.unesco.org/whc/index.htm

World Resources Institute (WRI)
http://www.wri.org/wri/

World Summit on Sustainable
Development (WSSD),
Johannesburg, South Africa,
September 2002
http://www.johannesburgsummit.org/

World Trade Organization (WTO)
http://www.wto.org/

World Water Council (WWC)
http://www.worldwatercouncil.org/

World Wide Fund
for Nature (WWF)
http://www.panda.org/

WorldWatch Institute
http://www.worldwatch.org

Index